RGS-IBG Book Series

The *Royal Geographical Society (with the Institute of British Geographers) Book Series* provides a forum for scholarly monographs and edited collections of academic papers at the leading edge of research in human and physical geography. The volumes are intended to make significant contributions to the field in which they lie, and to be written in a manner accessible to the wider community of academic geographers. Some volumes will disseminate current geographical research reported at conferences or sessions convened by Research Groups of the Society. Some will be edited or authored by scholars from beyond the UK. All are designed to have an international readership and to both reflect and stimulate the best current research within geography.

The books will stand out in terms of:

- the quality of research
- their contribution to their research field
- their likelihood to stimulate other research
- being scholarly but accessible.

For series guides go to http://www.blackwellpublishing.com/pdf/rgsibg.pdf

Published

Geomorphological Processes and Landscape Change: Britain in the Last 1000 Years
Edited by David L. Higgitt and E. Mark Lee

Globalizing South China
Carolyn L. Cartier

Lost Geographies of Power
John Allen

Geographies of British Modernity
Edited by David Gilbert, David Matless and Brian Short

A New Deal for Transport?
Edited by Iain Docherty and Jon Shaw

Military Geographies
Rachel Woodward

Geographies and Moralities
David Smith and Roger Lee

Forthcoming

Domicile and Diaspora
Alison Blunt

Fieldwork
Simon Naylor

Putting Workfare in Place
Peter Sunley, Ron Martin and Corinne Nativel

Natural Resources in Eastern Europe
Chad Staddon

The Geomorphology of Upland Peat
Martin Evans and Jeff Warburton

A New Deal for Transport?

The UK's struggle with the sustainable transport agenda

Edited by
Iain Docherty and Jon Shaw

Blackwell
Publishing

350 Main Street, Malden, MA 02148-5020, USA
108 Cowley Road, Oxford OX4 1JF, UK
550 Swanston Street, Carlton, Victoria 3053, Australia

First published 2003 by Blackwell Publishing Ltd
Reprinted 2004

Library of Congress Cataloging-in-Publication Data
A new deal for transport?: the UK's struggle with the sustainable transport agenda / edited by Iain Docherty and Jon Shaw.
    p. cm. – (RGS-IBG book series)
Includes bibliographical references and index.
    ISBN 1-4051-0630-1 (alk. paper) – ISBN 1-4051-0631-X (alk. paper)
    1. Transportation and state–Great Britain.
    2. Transportation–Environmental aspects–Great Britain. I. Docherty, Iain. II. Shaw, Jon. III. Series.

HE243.A2 N48 2003
388′.0941–dc21   2003000666

A catalogue record for this title is available from the British Library.

Set in 10/12 pt Plantin
by Kolam Information Services Pvt. Ltd, Pondicherry, India
Printed and bound in the United Kingdom
by MPG Books Ltd, Bodmin, Cornwall

For further information on
Blackwell Publishing, visit our website:
http://www.blackwellpublishing.com

To our families, and to Brian Hoyle, Richard Knowles and other TGRG colleagues who have supported and enriched transport geography over the years.

Rt Hon Jim Hacker MP, Minister for Administrative Affairs   'And after all, we do need a transport policy.'

Sir Humphrey Appleby, Permanent Secretary   'If by "we" you mean Britain that is perfectly true, but if by "we" you mean me and you and this department, we need a transport policy like an aperture in the cranial cavity.'

BBC (1982) *Yes, Minister: The Bed of Nails*, 9 December.

Contents

Series Editors' Preface

The RGS/IBG Book series publishes the highest quality of research and scholarship across the broad disciplinary spectrum of geography. Addressing the vibrant agenda of theoretical debates and issues that characterise the contemporary discipline, contributions will provide a synthesis of research, teaching, theory and practice that both reflects and stimulates cutting edge research. The Series seeks to engage an international readership through the provision of scholarly, vivid and accessible texts.

Nick Henry and Jon Sadler
RGS-IBG Book Series Editors

Notes on Contributors

Iain Docherty is a Research Fellow in the Department of Urban Studies at the University of Glasgow, and Secretary of the Transport Geography Research Group of the Royal Geographical Society (with the Institute of British Geographers). His research focuses on the impacts of political systems and structures of government on urban policy and city development strategies. Iain has written widely on many aspects of urban governance, including transport policy, and is the author of *Making Tracks: the Politics of Local Rail Transport* (Ashgate, 1999).

Jon Shaw is a Lecturer in the Department of Geography and Environment at the University of Aberdeen, and Chair of the Transport Geography Research Group of the Royal Geographical Society (with the Institute of British Geographers). He is also a member of the International Editorial Board of the *Journal of Transport Geography*. Jon researches and writes on a range of transport issues, including roads policy, rail privatization and social exclusion in rural areas. His previous books are *All Change: British Railway Privatisation* (McGraw-Hill, 2000) and *Competition, Regulation and the Privatisation of British Rail* (Ashgate, 2000).

John Farrington is Professor of Transport and Environment in the Department of Geography and Environment at the University of Aberdeen. His research interests focus on rural transport and accessibility, and on related issues of social justice and sustainability. He has led numerous research projects including an EU project on Environmental Transport Policy and Rural Development, and a UK Treasury funded project on Settlements, Services and Access in Rural Britain. He is the author of *Life on the Lines* (Moorland, 1984) and has published many articles on transport, sustainability and environmental assessment.

Phil Goodwin is Professor of Transport Policy at University College London, and Director of the Economic and Social Research Council's designated research centre on transport. He was formerly Head of Transport Studies at Oxford University, and an official of the Greater London Council. He has carried out research on travel demand and transport appraisal, was co-author of three reports by the Standing Advisory Committee on Trunk Road Assessment (on the environment, induced traffic and the economy) and has advised local and national bodies in many countries as well as the European Commission. He was chair of the advisory panel for the 1998 White Paper *A New Deal for Transport: Better for Everyone*, and is a nonexecutive director of the Port of Dover.

Brian Graham is a Professor at the University of Ulster and a former Chair of the Transport Geography Research Group of the Royal Geographical Society (with the Institute of British Geographers). He is also Advisor to the Department for Regional Development, Northern Ireland, on aviation matters. His research interests in air transport are concerned with the role of aviation in regional development and on the interaction between airline strategies and broader socioeconomic policies. He is the author of *A Geography of Air Transport* (John Wiley, 1995), co-editor of Ashgate's *Transport and Mobility* series and has published numerous papers on air transport in the UK and European Union.

Richard Knowles is a Reader in Geography in the School of Environment and Life Sciences at the University of Salford. His research interests and publications focus on transport deregulation and privatization, and on assessing the impacts of new transport infrastructure. He was a lead member of the Manchester Metrolink Light Rail Impact Study, cosponsored by the Economic and Social Research Council and Greater Manchester Passenger Transport Executive. He is the editor of the international quarterly research journal, *Journal of Transport Geography*, and co-editor of the textbook *Modern Transport Geography* (John Wiley, 1992, 1998).

John Preston is the Director of the Transport Studies Unit and Reader in Transport Studies at the University of Oxford. He is also the Tutorial Fellow in Geography at St Anne's College. His research in transport covers demand and cost modelling, regulatory studies, land-use, economic development and environment interactions. He has held over 60 research grants and contracts and has published over 100 articles, book chapters, and conference and working papers. His teaching and research has covered all modes of passenger and freight transport but with a particular emphasis on public passenger transport.

Austin Smyth is the Director General of Ireland's National Institute for Transport and Logistics. He is also Professor of Transport Economics at the Transport Research Institute at Napier University, and Director of the Transport Research Institute – Northern Ireland Centre. In 1989 he became the first Professor in the field of transport on the island of Ireland. He has served as an advisor on transport policy to devolved administrations in the UK, as well as the European Commission. Austin has worked extensively in and for the UK rail industry, and has consultancy experience across the European Union and in Eastern Europe, Russia, North America, the Middle East and Thailand.

Dominic Stead is a Senior Researcher at OTB Research Institute for Housing, Urban and Mobility Studies within Delft University of Technology. His research over the last 12 years has focused on the relationships between transport, land-use and environmental planning. During this time he has worked on a range of consultancy and research projects, which have been reported in conference papers, journals and books. He is co-author of *European Transport Policy and Sustainable Development* (Routledge, 2000).

Rodney Tolley is a Reader in Geography at Staffordshire University, where he is Director of CAST, the Centre for Alternative and Sustainable Transport. He has edited two editions of *The Greening of Urban Transport: Planning for Walking and Cycling in Western Cities* (Belhaven, 1990; John Wiley, 1997), which has been described as the 'bible' of green mode planning. He served as specialist technical advisor to the UK Government's inquiry into walking in 2001, and has convened and chaired the National Walking Conference in Britain since 1997. He is also the Director of the global partnership *Walk21*.

Geoff Vigar is a Lecturer in Spatial Planning and Transportation at the University of Newcastle upon Tyne. His research focuses on strategy development and implementation in land-use and transport planning, and the role and incorporation of social and environmental issues into governance practices in particular. He is the author of *The Politics of Mobility: Transport, the Environment and Public Policy* (Spon, 2002) and co-author of *Planning, Governance and Spatial Strategy in Britain* (Macmillan, 2000).

William Walton is a Senior Lecturer in the Department of Geography and Environment at the University of Aberdeen. His research interests include the relationship between road building policy and land-use planning. He has written numerous journal articles on topics including roads policy in the UK at the national and the local level, the deregulation of the provision of motorway service areas, air pollution, traffic impact analysis and multi-modal studies.

Peter White is Professor of Public Transport Systems at the University of Westminster, where he is responsible for postgraduate teaching and research in bus, coach and rail systems. He is the author of *Public Transport: its Planning, Management and Operation* (Spon, 2002) and numerous published papers. His particular interests in recent years have focused on the effects of privatization and deregulation in the coach and local bus industries, and the impact of rail privatization in Great Britain. He is currently responsible for managing the University of Westminster's contribution to the joint study updating the TRL handbook, *The Demand for Public Transport*.

Foreword

Professor David Begg
Chair, Commission for Integrated Transport

It's much easier being an academic commenting on the government's track record on transport than it is to be the politician at the sharp end making the tough decisions. Having been the lead politician on transport in Edinburgh for five years, pursuing an ambitious and radical sustainable transport agenda, I have personal experience of how tough it can be. You will therefore excuse me if I am less critical than some of my academic colleagues who have written chapters in this book when commenting on the government's performance.

If it is any consolation to past and present transport ministers, I can think of few reviews of government performance on transport which have been anything other than extremely critical. The one exception in my lifetime was the commentary on the 1998 White Paper, *A New Deal for Transport*, which, at the time, was met with widespread support. The White Paper offered the prospect of changing the way we travelled in the UK with less dependence on the car. With John Prescott at the helm, there was a window of opportunity to get the Whitehall machine behind the radical new agenda. There has been no other occasion when someone as senior as the Deputy Prime Minister has held the transport brief.

To my mind, the most radical policy in the White Paper was the proposal to give local authorities and the Mayor of London the power to introduce congestion charging or workplace parking levies, with the revenues raised ring-fenced for public transport improvements. The concept of charging had been championed in academic and local authority circles for more than a decade but the potential political fall out from introducing 'new taxes' on motoring together with opposition from the Treasury to the concept of 'hypothecation', meant that it never really became a serious policy proposal. This changed with the arrival of the White Paper and subsequent legislation in the Transport Act (2000)[1] which paved the way for the Central London Congestion Charging Scheme, the most radical attempt to change travel behaviour and reduce traffic congestion by fiscal means anywhere in the democratic world. As this book is being printed the £5 congestion charge to enter the centre of London has recently gone live.

Yes, there have been forms of road pricing in operation in the Norwegian cities of Oslo, Bergen and Trondheim for more than a decade. But there the main objective is to raise revenue to finance transport infrastructure and not primarily as a congestion reduction measure which is the case with the London scheme. Yes, an area cordon congestion charging scheme has been in operation in Singapore since 1975 with a more technical sophisticated electronic road pricing scheme since 1996. But Singapore is not a democracy. The London congestion charge scheme is pioneering and the eyes of the world will be watching closely. The Labour government deserves credit for doing what no previous government has been willing to contemplate since congestion charging was first recommended by the Smeed Committee in 1964.

The central contention in this book is that since 1997 the government has lacked radicalism in its attempts to reduce car dependency. I would argue that this is true for most governments in the democratic world where there is reluctance to be viewed as 'anti-car'. The UK government deserves credit for legislating to give enabling powers to local authorities to introduce congestion charging and to hypothecate the revenue. What they have failed to do is to provide leadership to encourage any take up. Without the spread of congestion charging from London out to other towns and cities in the UK, it will be impossible for the government to achieve its own congestion reduction target by 2010. The mistake was to have placed too much of a burden on local authorities without providing any underpinning leadership and support from central government. The difficulties local authorities would face in introducing charges were also underestimated.

The Cabinet did not give John Prescott the financial support he needed between 1997 and 1999 when expenditure on transport fell below the level of the previous Conservative government. That has certainly altered recently by what can only be described as a step change in public expenditure on transport. Between 2000/01 and 2003/04, government spend will have almost doubled in real terms from £4.8 billion to £9.3 billion. Yet the extra expenditure has actually given rise to other problems. First, there is now less incentive to introduce fiscal demand management measures to raise finance locally for transport spending. Second, there has been ineffective use of the additional funds by the two delivery 'agencies' that are central to the sustainable transport agenda: areas of local government and areas of the rail industry.

The growing demand from local government through the 1990s to be given powers to introduce congestion charging was driven more by their desperate need for new sources of finance than it was to change travel behaviour. With Local Authority capital allocations doubling between 2001 and 2003, evidence is currently being gathered which shows that some are not able to spend the money allocated and that the Single Capital Pot has resulted in 'transport money' being spent on other services instead. Local authorities – certainly in England – have been able to fund the 'carrot' without resorting to the 'stick' to raise finance. The government's decision to fund the Leeds Supertram without making congestion charging an integral part of the deal was an error of judgement.

There are frequent criticisms of the government that it is starving the railways of finance. This is not my interpretation of the statistics. Government expenditure on rail is forecast to increase by 30 per cent between 2001/02 and 2002/03. Most Whitehall departments would be envious of such a percentage increase. But because of the dramatic fall in rail productivity, the extra money has been able to purchase less and less. This is why the projected cost of track maintenance and renewal to 2011 has doubled from the £20 billion originally estimated by the Rail Regulator, Tom Winsor. The Chairman of the SRA, Richard Bowker, is absolutely right to place so much emphasis and concentrate so much of his effort on reducing the cost of operation, renewal and enhancements. I can always put a passionate case for more expenditure on transport, especially if it results in making alternatives to the car more attractive. But history tells us that if there is a squeeze on public expenditure – not a wholly unlikely prospect – then transport is always vulnerable. Unless rail productivity improves, particularly the return from rising expenditure on track infrastructure, then the case for continuing the recent increase in public spend is very fragile indeed.

The government is pursuing a large road building programme as part of its transport strategy, but to argue – as some do – that ministers are favouring road expenditure over rail, would be wrong. Public investment in rail over this decade is forecast to more than double compared with the 10 years previous, from £6.4 billion to £14.7 billion. By contrast, investment in roads over the same period rose from £11.5 billion to £13.6 billion today. Rail is also revealed as the clear winner in the recommendations made to the government in its programme of multi-modal studies – which suggest that over half of the total spend (55 per cent) should be directed to the railways, 28 per cent to road infrastructure and 17 per cent to Local Transport Capital – although most of the rail projects recommended are unlikely to be delivered this decade because of Strategic Rail Authority funding constraints.

So has there been a U-turn in government transport policy since 1997? The emphasis has certainly changed. In 1997/98 the focus was very much on reducing the need to travel, achieving a modal switch from car to public transport, walking and cycling and reducing car dependency. As John Prescott said, 'we have to make hard choices on how to combat congestion and pollution while persuading people to use their cars a little less'.[2] These policy objectives require a level of traffic restraint which the government has been unwilling to embrace since the fuel duty escalator was abolished in September 2000. The main policy objective at present seems to be to cater for the increased mobility of the population whether by road, rail or air which has resulted from a growing economy. Is it still the government's aim to reduce the need to travel as well as improving choice? If it is not, as this book suggests, then there has most certainly been a policy shift since the 1998 White Paper which I, for one, could not support.

NOTES

1 Transport Act (2000) *Public general Acts – Elizabeth II*. Chapter 38. The Stationery Office, London.
2 Department of the Environment, Transport and the Regions (1988) *A new deal for transport: better for everyone*. Cmnd 3950. The Stationery Office, London, 3.

Preface

The aim of this book is simple – to assess the government's record on transport. We should state at the outset that we generally agree with the analysis Labour made of the transport problem before coming to power. The Party's position, in essence, rejected the myth of the 'great car economy'. It held that contemporary trends in transport were unsustainable. Established levels of traffic growth (and perhaps even the current level of traffic) could not continue, because increasing congestion and pollution constrain the economy, harm quality of life and threaten the environment. Public transport provision, substandard after years of neglect and underinvestment and further compromised by the Conservatives' dogmatic structural reforms, would have to improve. People should walk and cycle more. There was an overwhelming acknowledgement that 'things could not go on as they have before', and this view was shared by many in the transport community.

We contend that whilst Labour's original policy *intent* to promote sustainable transport was welcome, desirable policy *outcomes* have to date been minimal. As such, the time has come to ask whether the heralded *New Deal for Transport* is a chimera. The evidence suggests a powerful case against the government: car use has continued to rise; large-scale road building has crept back up the agenda; plans for road user charging remain absent in the vast majority of the country; the further expansion of train travel is under threat; a large increase in airport capacity is likely; and the promotion of other modes such as buses and cycling has been less than emphatic. There are creditable exceptions – and these are discussed in the pages that follow – but given the urgency Labour originally attached to addressing the UK's transport problems, its performance to date has been disappointing.

In advancing our principal argument, we are aware of some mitigating factors. Bringing about major changes in the transport sector is a long-term business. It would be unrealistic, even six years after Labour came to power, to expect a large number of major infrastructure improvements to have come to fruition. Yet genuine and considerable progress in terms of making things better for the future should be well underway. In addition to primary legislation being enacted swiftly and a sufficient amount of capital being made available, many small-to-medium scale improvements to the transport network, such as new and substantially expanded bus and train fleets, additional cycle paths and improved pedestrian access, should now be a reality. With regard to larger projects, upgrade plans should have progressed to an advanced stage and, where appropriate permissions have been granted, construction work should be proceeding apace. In other words, where stated policy outcomes have not yet been achieved, their delivery should be in progress.

There are also issues associated with the gap between political rhetoric and the realities of the policy process. How much can politicians reasonably be expected to deliver on pledges given in the heat of an election campaign? Is it disingenuous to judge a government's achievements over a five or six year period in terms of its original commitments given the inevitable need for policy to shift in reaction to, or anticipation of, changing circumstances? There is good reason to believe that senior figures in the incoming administration were genuinely committed to making transport in the UK more sustainable – most of Labour's original policy intent was articulated in official Party or Whitehall documentation before the publication of what we and many others saw as a rather watered-down White Paper in 1998 – but if ministers were surprised by the scale of the political difficulties which arose as they sought to make the 'hard choices' of government, then they alone must shoulder the blame for not having foreseen this.[1] Promoting sustainable transport was never likely to be easy in the UK's car dependent society, and it is scarcely credible to believe that the government-in-waiting viewed the prospect of, say, appreciably increasing petrol duty as anything other than politically damaging. Labour had, after all, been fatally wounded by the public's self-interest in resisting proposed tax increases in the 1992 general election.

So whilst we accept that politicians are not usually given to intentionally jeopardising their electoral prospects, the sheer strength of the government's mandate – majorities of 179 and 167 in the 1997 and 2001 general elections – might, *should*, have provided enough 'air cover' for them to take at least some of the unpopular decisions necessary to advance the cause of sustainable transport. In short, the fundamentals of Britain's transport problems have not changed since 1997. Things in 2003 still cannot 'go on as they have before', and ministerial attempts to redefine or back away from Labour's previously stated intentions warrant critical examination.

The book begins by setting out the political context in which UK transport policy is made, administered and implemented. Against this background, a series of chapters reviews progress in relation to each of the main transport modes. A discussion of prospects for the future brings the volume to a close. We have tried to ensure that the book has been structured in a 'user-friendly' way. Those who wish to read its contents from cover to cover should find that the text tells a logical story. Equally, readers who consult only one or a handful of chapters should find that each contribution mostly holds its own as a stand-alone piece. Where reference to other chapters for relevant contextual information is necessary, appropriate signposting is provided.

Although the book has been produced on behalf of the Transport Geography Research Group of the Royal Geographical Society (with the Institute of British Geographers), we have assembled bespoke contributions from a wide range of authors from different subdisciplines within the transport field. In passing judgement on Labour's transport record since 1997, geographers are joined by transport academics from a number of cognisant disciplines, including economics, political science, policy analysis and strategic planning. The book benefits from the years of collective experience and expertise of its contributors, but we would not claim to have put together the definitive analysis of our topic. Rather, we hope that the volume will make a positive contribution to the ongoing and recently re-energized transport debate. Each author has presented his arguments clearly and underpinned them with rigorous and transparent justification, but readers should decide for themselves on the validity of the judgements.

As editors we owe debts of gratitude to many. We would first and foremost like to thank our fellow contributors. They have all provided extremely well-written chapters in a timely fashion and accommodated our various comments on their original manuscripts. Ian Bailey, Steve Bennett, Clive Charlton, Nigel Harris and Christian Wolmar cast their eyes over various parts of the text and brought our attention to the usual editorial mistakes and omissions. Alison Sandison and Jenny Johnston turned the figures – which they received in various formats and levels of decipherability – into handsome, consistently styled artwork. Finally, Angela Cohen and Debbie Seymour at Blackwell's and our RGS-IBG series editor Nick Henry provided very welcome support, encouragement and advice throughout the project, as did our colleagues in the TGRG. Obviously the usual disclaimer applies, but we hope that all of the above are happy with the final product.

On a personal level, our friends and families have helped keep us sane over recent and increasingly frenetic months as we finalized the text. Particular thanks are due to a small bunch in Scotland's first and third cities. In Glasgow, Andrea kept checking to make sure Iain was busy, whilst Stuart Gulliver, Elaine Bence and colleagues from the Bute Gardens labyrinth reminded both of us to have the courage of our convictions. Derek

Hall provided the ideal companion with whom to discuss the finer points of policy, politics and pakora. In Aberdeen, the Yellow Corridor Posse – Danny MacKinnon, Doug Mair, Andy McMullen and Coll – livened up Jon's workplace and, with Tim Curtis (for a weekend in November), Murray Grant, Iain Malcolm and others, the granite city beyond it. Finally, Roland Gehrels and Shea Meehan's 'out-of-town' linguistic advice proved highly significant, as well they knew it would.

Now we promise we'll stop going on about 'the book'.

Iain Docherty,   Jon Shaw,
Glasgow, Scotland   Bromley, England

                                  February 2003.

NOTE

1   Department of the Environment, Transport and the Regions (1998) *A new deal for transport: better for everyone.* Cmnd 3950, The Stationery Office, London, 3.

Abbreviations

AM	Assembly Member (Wales)
BA	British Airways
BBC	British Broadcasting Corporation
BIC	British-Irish Council
BNRR	Birmingham Northern Relief Road
BR	British Rail
BSOG	Bus Service Operators' Grant
CC	Competition Commission
CBI	Confederation of British Industry
CDG	Paris Charles de Gaulle Airport
CfIT	Commission for Integrated Transport
CoBA	Cost Benefit Analysis
CPRE	Council for the Protection of Rural England
CTRL	Channel Tunnel Rail Link
DBFO	Design, Build, Finance and Operate
DEFRA	Department of the Environment, Food and Regional Affairs (UK)
DEMU	Diesel-Electric Multiple Unit
DETR	Department of the Environment, Transport and the Regions (UK)
DfT	Department for Transport (UK)
DLR	Docklands Light Railway
DMU	Diesel Multiple Unit
DoE	Department of the Environment (UK)
DoT	Department of Transport (UK)
DRD	Department of Regional Development (Northern Ireland)
DTI	Department of Trade and Industry (UK)

DTLR	Department of Transport, Local Government and the Regions (UK)
EA	Environmental Assessment
ECMT	European Conference of Ministers of Transport
EEA	European Economic Area
EMU	Electric Multiple Unit
EU	European Union
FDR	Fuel Duty Rebate
FFG	Freight Facilities Grant
GDP	Gross Domestic Product
GLA	Greater London Authority
GOR	Government Offices for the Regions (England)
ICAO	International Civil Aviation Organisation
IPPR	Institute for Public Policy Research
JLE	Jubilee Line Extension
JMC	Joint Ministerial Committee
KM	Kilometre
LRT	Light Rapid Transit
LTP	Local Transport Plan
LTS	Local Transport Strategy
LUL	London Underground Limited
MCC	Metropolitan County Council
MLA	Member of Local Assembly (Northern Ireland)
MMS	Multi-Modal Study
MP	Member of Parliament (UK)
MPH	Miles Per Hour
MSP	Member of the Scottish Parliament
NAO	National Audit Office
NATA	New Approach to Appraisal
NCS	National Cycling Strategy
NIR	Northern Ireland Railways
NRTF	National Road Traffic Forecasts
NWS	National Walking Strategy
ODPM	Office of the Deputy Prime Minister (UK)
OECD	Organisation for Economic Cooperation and Development
OEF	Oxford Economic Forecasting
OFT	Office of Fair Trading
ONS	Office for National Statistics
OPRAF	Office of Passenger Rail Franchising
ORR	Office of the Rail Regulator
PFI	Private Finance Initiative
PPG	Planning Policy Guidance

PPP	Public Private Partnership
PRN	Primary Route Network
PSBR	Public Sector Borrowing Requirement
PSO	Public Service Obligation
PTA	Passenger Transport Authority
PTE	Passenger Transport Executive
RASCO	Regional Air Studies Coordination
RRC	RASCO Reference Case
RCEP	Royal Commission on Environmental Pollution
RDA	Regional Development Agency (England)
ROSCO	Rolling Stock Company
RPG	Regional Planning Guidance
RSS	Regional Spatial Strategy
RTS	Regional Transport Strategy
SACTRA	Standing Advisory Committee on Trunk Road Assessment
SEC	South East Constrained
SERAS	South East and East of England Study
SPV	Special Purpose Vehicle
SRA	Strategic Rail Authority
T5	Terminal Five, Heathrow Airport
TfL	Transport for London
TOC	Train Operating Company
TPP	Transport Policies and Programmes
UDP	Unitary Development Plan
UK	United Kingdom
UKC	UK-Wide Constrained
UN	United Nations
WCML	West Coast Main Line

Part I

Policy and Politics

1

Policy, Politics and Sustainable Transport: The Nature of Labour's Dilemma

Iain Docherty

The 1997 Labour government promised to introduce radical transport policies aspiring to the goal of much-improved economic, environmental and social sustainability. Central to this was the desire to build on the 'Consensus for Change' in transport policy identified by the party while in opposition in the mid-1990s.[1] This consensus was built around the recognition that past policies aimed at accommodating relentless increases in the demand for travel were failing. Reducing the dominance of the car would be essential since a deepening transport 'crisis' was developing.[2] Increasing congestion and unreliability of transport services undermined the sustainability of the economy; transport-related pollution, deteriorating local air quality and greenhouse gas emissions, threatened the sustainability of the environment; and unequal access to transport contributed to the problems of social exclusion that jeopardized the sustainability of many local communities.

Despite its promises to make 'hard choices' in transport policy, Labour has struggled to implement sustainable transport strategies in government, and to convince the public of their value.[3] This chapter provides an overview of Labour's record and its stated plans for the future, analysing why it has been unable to live up to its own aspirations to radically change the direction of transport policy. It charts a series of key events in the government's first six years, such as the launch of the 1998 White Paper, *A New Deal for Transport: Better for Everyone*, and the fuel tax protests of 2000, and addresses ministers' fundamental unwillingness to implement the changes required to enhance the sustainability of the economy, the environment and society.[4]

Historical Context

The increased flexibility and individual choice of when and where to travel associated with widespread car ownership and use has transformed almost every aspect of our society. As access to cars has increased, people have travelled further between their homes, workplaces and places of consumption. The urban decentralization and deconcentration of the post-war era have also made these patterns more complex, as trunk radial flows of movement to and from major urban centres have been supplemented by a complex web of circumferential and tangential trips.

Over time, changing patterns of land use have reflected the widening availability of transport, and its increased effectiveness in reducing the friction of distance. Before 1800, the land transport systems that provided the means of economic exchange between settlements were exclusively based on roads for the use of pedestrians and horse-drawn vehicles. However:

> Major changes in transport technology which began to emerge during the first quarter of the 19th century had a major influence on the growth of cities, the organization of their internal structure, and the supply, demand, efficiency, speed and opportunities for movement within them.[5]

The subsequent shift from 'foot cities' to 'tracked cities' had profound implications for settlement form.[6] Land uses became increasingly separated and specialized, and technological advances, such as the development of tram and metro networks in the early twentieth century, further encouraged urban dispersal. The suburb, built at much lower residential density than the historic areas of the inner urban core, became the aspirational choice of residence for the majority.

But the 'tracked' era was to last little over 50 years. After 1920, transport in the UK, as in most countries of the developed world, was transformed by the introduction of motor vehicles, particularly the private car. Although it is widely perceived that the political rhetoric underlying the promotion of widespread car ownership came from the Right – for whom the car was crucial to both personal liberty and the promotion of flexible, responsive markets – it is important to recognize that the Left also has a long tradition of regarding increased car ownership and use as desirable.[7] At the heart of this position is a utopian vision of the economy and society, which incorporated universal car ownership as a solution to the transport equity dilemma of unequal access to travel, and the opportunities for employment and consumption that it creates, between social groups.[8] As Frank Lloyd Wright, one of the most eloquent advocates of this vision, eulogized:

> What nobler agent has culture or civilization than the great open road made beautiful and safe for continually flowing traffic, a harmonious part of a great whole life? Along these grand roads as through human veins and arteries throngs city life, always building, building, planning, working.[9]

Indeed, such was the importance attached by the Left to the accommodation of the private motor vehicle within overall urban transport policy that the origins of the notion of 'predict and provide' – the idea that the amount of road space should be expanded as far as possible to meet the demand for car travel – can be traced back to the celebrated prewar socialists Beatrice and Sidney Webb, who said:

> we cannot doubt that – whatever precautions may be imposed for the protection of foot-passengers, and whatever constitutional and financial readjustments may be necessary as between tramways, omnibuses and public revenues – the roads have once more got to be made to accommodate the traffic, not the traffic constrained to suit the roads.[10]

After 1945, Britain embarked on a significant road building programme designed to support the regeneration of the economy. Whereas before the war, strategies had focused on piecemeal upgrading of existing major routes, there was now the opportunity to greatly increase the scale and ambition of the strategy. Inspired by the freeways and parkways of North America, strategic regional plans, such as Sir Patrick Abercrombie's *Greater London Plan* and *Clyde Valley Regional Plan*, envisaged a dense network of express roads around each major city.[11] A programme for the construction of a national interurban motorway network was also drawn up, with the first section, a bypass of Preston, Lancashire, opening in 1958.

As the national road building plan was gaining momentum in the early 1960s, the then government published a seminal document which crystallized debate on what a car dominated future would look like. *Traffic in Towns*, better known as the 'Buchanan Report' after its author, the late Sir Colin Buchanan, envisaged the changes in the physical structure of British towns and cities required if they were to adapt to accommodate unrestricted use of the car. Although the report was much vilified at the time as representing 'motorway madness', its core message was that severe congestion was the inevitable outcome of the failure to match increased supply of road space to the voracious appetite for car travel.[12] Put simply, the government could either try to predict and provide – build sufficient new road space to match the forecast increase in car traffic – or find alternatives to unrestricted car-based mobility. Yet Buchanan was also the first to identify how a 'car-owning democracy' had emerged:

> It seems futile to deny these things [the advantages of motorcars]. The motor vehicle is a remarkable invention, so desirable that it has wound itself inextricably into a large part of our affairs. There cannot be any going back on it.[13]

The importance of the concept of the car-owning democracy is that it neatly summarized how the demand for personal mobility was likely to be

insatiable as more and more people's lives were transformed by the possibilities offered to them by the car. However, the UK's adoption of the
North American 'orthodoxy' of universal car ownership and use has
become every bit as problematic as Buchanan (and others) feared it
might.[14] Britain's towns and cities have followed the American trends
towards low density suburban sprawl and the rapid growth of satellite
dormitory settlements around major cities, encouraged by a *laissez-faire*
attitude to widespread car use.[15] A vicious circle was created, as people
tried to escape the congestion and declining quality of life of major urban
centres by commuting ever-greater distances.

These trends in land use and transport then reinforced each other over
several decades, resulting in a situation of widespread 'car dependence'.
Many people, particularly those locked into sub- and ex-urban land-use
patterns, now require (very) high levels of mobility simply to maintain their
lifestyles.[16] For example, a major MORI/BBC survey in 1999 found that
fully 79 per cent of drivers agreed that it would be difficult to adjust their
lifestyles to being without a car.[17] At the same time, people without access
to a car find their situation deteriorating, as public transport provision
declines in response to reduced demand and the shift of major activities,
such as employment, leisure and retailing, to sites on the urban fringe that
are difficult to access without a car. The result is social exclusion, or the
erosion of social sustainability, as the American writers K. H. Schaeffer and
Elliott Sclar explained almost 30 years ago:

> It is our contention that the urban crises which manifest themselves in so
> many ways have at least one common root. This is the increasing reliance
> on the automobile. In every urban area, the automobile has become the
> only means of transportation by which every part of the region can be
> reached . . . Wherever the automobile is the mode of travel, there access to
> transportation is distributed very unevenly between individuals. This is prob-
> ably the greatest social fault of the automobile.[18]

That both Buchanan's prediction that the increase the supply of road
infrastructure would fall far short of meeting rising demand, and Schaeffer
and Sclar's bleak vision of urban economic and social decay, have come
true underlines the extent of the 'travel sickness' afflicting modern Britain.[19] The depth of this malaise is in part explained by the state's late
realization that it could not simply build its way out of congestion. Despite
the elusiveness of predict and provide – which was neither a desirable (in
terms of sustainability) nor indeed feasible basis on which to construct a
strategy for transport – successive post-war administrations tried hard to
prove Buchanan wrong by attempting to expand road space as much as
possible.

Until the early 1990s, the core concern of roads policy (and by implication transport policy more generally) in Britain remained the straightforward implementation of predict and provide. In many ways, the policy
process was really quite simple. The basic premise was that rising standards
of living necessitated increased car ownership and use. This trend was well
established and showed no signs of changing (Figure 1.1). Moreover,
greater car-based mobility was seen to both enhance individual liberty
and boost the economy – directly through the growth of the motor industry,
but also more generally since increased physical mobility helped to liberalize housing and labour markets. It was therefore deemed essential to
accommodate as much car use as possible, or, as Phil Goodwin neatly
summarized, since 'private car use would increase . . . it was necessary to
increase roads capacity. And public transport use would decline, therefore
it would be logical to reduce service levels'.[20] Gestures towards the goals of
better social and environmental sustainability were largely limited to the
maintenance of some 'lifeline' public transport services in fragile communities and attempts to mitigate against the impacts of car traffic on local air
quality, such as the move to unleaded petrol.

By the end of the 1980s, the combination of the Thatcher government's
support for the car owning democracy and the Lawson economic boom set
the scene for the pinnacle of predict and provide with the publication of the

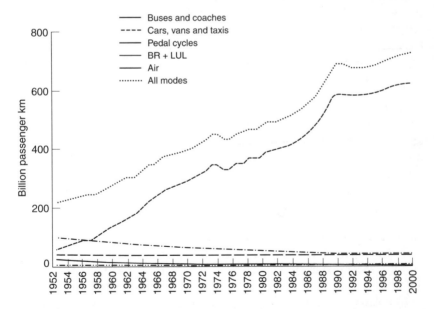

Figure 1.1 Passenger transport by mode, 1952–2000. Source: Department of Transport, Local
Government and the Regions (2001) *Transport Statistics Great Britain: 2001 Edition*. DTLR, London.

White Paper, *Roads for Prosperity*.[21] The White Paper's 494 road schemes in England would have cost more than £12 billion. With the addition of maintenance and minor improvements to the existing network, the total package represented an investment of some £23 billion (over £33 billion at 2003 prices), and was championed by ministers as 'the largest road building programme since the Romans'. Yet this was 10 years after the OECD (an organization not noted for advocating radical state intervention in the market) warned that a strategy focused on road building would be unlikely to solve the transport problem:

> since further extension of the road infrastructure to meet growing demand for car use is not everywhere possible for urban planning and financial reasons, nor desirable from environmental, energy and often social policy standpoints, the only remaining transport policy option is to swing modal split in favour of public transport by investment and/or pricing policy measures.[22]

Sustainable Transport as a Political Issue

In retrospect, it is highly ironic that the whole discourse surrounding transport policy in the UK was to change fundamentally almost as soon as *Roads for Prosperity* was published. As in many other areas of politics, it was unforeseen 'events' of the kind famously bemoaned by Harold Macmillan that were to disrupt the *status quo* of British transport policy. The first such 'event' was probably the reaction to the publication in 1987 by the UN Commission on Environment and Development of a far-reaching report on the future of the global environment. The report, *Our Common Future* (commonly known as the 'Bruntland Report' after the Commission's Chair, Gro Harlem Bruntland), for the first time set out the scale of the environmental problems that could arise if contemporary development trends were left unchecked, especially the voracious consumption of natural resources and increasing pollution of air, water and land. In a very well-known passage, the Commission offered a definition of sustainable development, which has since been widely adopted: 'sustainable development is development that meets the needs of the present without compromising the ability of future generations to meet their own needs'.[23]

In essence, it was the realization of the potential scale of any impending environmental crisis, and transport's contribution to any such crisis, that placed the word 'sustainability' in general usage and marked the 'turning point' or 'watershed' in transport policy.[24] As William Black put it when paraphrasing the Bruntland Commission, the challenge for transport was to achieve a sufficient level of sustainability that would 'satisfy current transport and mobility needs without compromising the ability of future generations to meet these needs'. The emerging concept of 'sustainable transport' was somewhat slippery and ill-defined, since, as Black continued, 'there is no limit placed on "future generations" and nothing is sustainable forever'.[25] Nevertheless, over the next few years, the imperative – at least at the rhetorical level – would become finding ways of minimising the environmental impacts of transport.

If any single event could be said to mark the beginnings of the search for a sustainable transport paradigm, then the 1989 meeting of the European Conference of Ministers of Transport (ECMT) seems the most likely candidate. Following the dynamic created by the 1987 World Commission Report, the 1989 ECMT received a number of scientific papers arguing that transport was an ever-increasing threat to environmental sustainability, particularly through the emission of greenhouse gases.[26] The message that the transport sector had become one of the most rapidly growing sources of global pollution was stark enough, but what was most striking was the acknowledgement that the majority of this pollution could be attributed to private traffic. The relentless increase in car use, which had already undermined the social sustainability of countless *local* communities, was now threatening to undermine the environmental sustainability of the *global* community. In other words, the environmental impact of ever-increasing traffic had made apparent the futility of predict and provide in a way that arguments over its usefulness as a planning strategy never had.

After a succession of further events, most importantly the UN *Earth Summit* held at Rio de Janeiro in 1992, the UK government was prompted into action and formally changed its transport policy. The realization of the scale of transport's environmental impact reinforced the value of long-standing policies of enhancing public transport and expanding opportunities for walking and cycling which had been prevalent across most of continental Europe for several decades. But in the UK, where an essentially North American model of car use had been prioritized, achieving the goals of sustainable transport would pose more fundamental challenges. In response to Rio, the 1992–7 Major government charged the Royal Commission on Environmental Pollution (RCEP) with advising ministers on how UK policies should be adapted to meet growing global environmental concerns. The Commission's eighteenth report, *Transport and the Environment*, restated the need for a fundamental change in the government's official stance on the future of transport policy in Britain:

> There is now general recognition that a continuing upward trend in road traffic would not be environmentally or socially acceptable. The need is to find transport policies for the UK and Europe which will be sustainable in the long term.[27]

After the publication of the RCEP report, the government transformed its official position almost overnight, ditching the last vestiges of predict and provide. Roads policy was now to be more about managing the car and its impacts rather than accommodating them, and transport policy was to be less about roads and more about a balance between modes. But however much advocates of sustainability claim the recognition of the environmental impacts of transport led the reappraisal of existing strategies, it is at least as likely that the move towards this approach to transport policy was inevitable even without increasing environmental concern.[28] The recession of the early 1990s reduced the resources available for road building, and underlined the impossibility of meeting the aspirations of plans like *Roads for Prosperity*, even if this were deemed desirable.[29] At the same time, popular protest against numerous road building schemes made delivering even quite modest new roads more time-consuming and expensive.

On assuming office in 1997, Labour began the process of delivering the 'Consensus for Change' it had identified whilst in opposition.[30] Within six weeks of taking power, Deputy Prime Minister John Prescott (whose responsibilities included those of Secretary of State for the Environment, Transport and the Regions) was characteristically bullish about the government's ability to implement a more sustainable agenda for transport. Faced with road traffic forecasts predicting 50 per cent growth in 30 years (Figure 1.2), he agreed that 'doing something about traffic' was essential since 'the forecast growth in traffic is clearly unacceptable'.[31] In a memorable statement, he demonstrated considerable belief in the government's (and his own) ability to deliver by saying: 'I will have failed if in five years time there are not many more people using public transport and far fewer journeys by car. It is a tall order but I urge you to hold me to it'.[32]

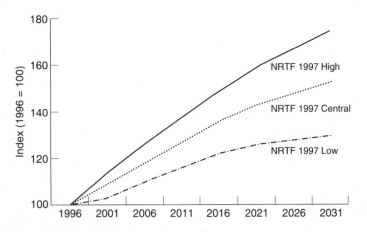

Figure 1.2   National road traffic forecasts (NRTF), 1997. Source: Department of the Environment, Transport and the Regions (1997) *National Road Traffic Forecasts (Great Britain) 1997*. DETR, London.

Despite offering such a 'hostage to fortune' (the targets were not attained), there appears little doubt that John Prescott's enthusiasm for, and belief in, the sustainability agenda was genuine.[33] In the government's early weeks in office, he instigated a number of radical initiatives. These included a complete review of the inherited trunk-roads programme utilising the New Approach to Appraisal (NATA), which was designed to incorporate a wider set of criteria than had been used previously to evaluate roads projects (Chapter 4). He also reiterated his commitment to introducing a much stronger direction to railway policy (which was widely seen to have 'drifted' since privatization), by proposing the creation of a Strategic Rail Authority (SRA) charged with developing the network (Chapter 5).

Within four months of taking power, Labour began to expand on its initial statements when it published a consultation document, *Developing an Integrated Transport Policy*,[34] which represented a 'dispassionate account of the problems as they appeared to the incoming Government'.[35] The problems identified were many and various, but can be grouped together under the three general concepts outlined in the introduction to this chapter: sustainability of the environment, sustainability of the economy and sustainability of society. On environmental sustainability, Labour acknowledged the growing global concern over climate change, and the work of the RCEP in informing the debate over how the environmental impacts of transport could be reduced. At the 1997 UN conference on climate change in Kyoto, one of its first major international summits, the new government (represented by Prescott) supported the adoption of the protocols committing developed nations to significantly reducing their carbon emissions to below 1990 levels by 2010. Since transport was estimated to account for around 25 per cent of the UK's emissions – with road vehicles accounting for four fifths of the transport total – this was widely seen at the time to be an important early signal that the new government was indeed prepared to act to reduce the environmental impact of transport generally, and road traffic in particular. Early policy statements reinforced this perception, with ministers floating a range of potential demand management measures including congestion charging, motorway tolls and workplace and retail parking taxes. Particular journeys, such as the 'school run' and supermarket trips, were highlighted as being especially amenable to modal shift. In its first Budget, the government even raised the fuel tax escalator – the additional annual increase in fuel duty above inflation introduced by the Conservatives on the advice of the RCEP – from five per cent to six per cent.

The sustainability of economic prosperity was highlighted as equally important by Labour. Keen to be portrayed as 'the party of business', the new approach to transport policy was presented as being in the interests of key producers. The economic cost of delays to traffic, estimated at around

£15 billion annually by the CBI, was highlighted, as was the importance of improving international links for export-led sectors increasingly at risk from unreliability of logistics chains in an era of 'just-in-time' deliveries.[36] Although not analysed in detail, there were also some references to the importance of transport infrastructure and provision in attracting inward investment, and to the role of public transport in larger towns and cities in sustaining their position as competitive locations for service and knowledge industries.[37] This was seen as particularly important in London, where the reduction of average travel speeds to near nonmotorized levels was perceived as a serious disincentive to economic development.

The third 'problem set' was that of the impact of transport externalities on social sustainability. Reflecting the desire to ensure 'joined-up government', some quite insightful statements were made on the impacts of transport patterns on health, social polarization and urban regeneration. One of the biggest criticisms of Conservative policy was that transport – and even each individual mode within the transport sector – was treated in isolation from interconnecting activities and policies. To counter this, Labour's early statements made much of two issues in particular, namely health and social exclusion. On health, the government encouraged a debate on the human and financial costs of transport-related pollution, estimated at more than £17 billion per annum, with an estimated 24,000 premature deaths per year linked to air pollution.[38] Statements on walking and cycling also linked these modes to the promotion of healthier lifestyles, since both modes have a clear role to play in preventative health care by encouraging people to undertake more physical activity.

On social exclusion, improving the availability and quality of public transport was deemed especially important, with the low levels of car access in particular social groups, including women, the elderly, the young and the unemployed, as being a major barrier to 'a fairer, more inclusive society'.[39] Proposals to extend access to excluded groups were undoubtedly weakened, however, by Labour's desire to keep the provision of public transport services in the private sector. None of the government's statements satisfactorily explained why private transport companies would (or should) be willing to alter market-driven service patterns and fares to suit the aspirations of policies aimed at reducing social exclusion.[40]

Labour's explicit recognition of the importance of economic and social as well as environmental policy aspirations aligns the Party's approach to transport with the oft-quoted 'three legged stool' conception of sustainability. This is clearly not unreasonable for an elected administration, since the 'eco-authoritarian' or 'deep green' standpoint elevates the protection of the environment above all other considerations including the maintenance of democracy and justice.[41] 'Light green' approaches to sustainability can be, in theory at least, very effective provided that policies designed to safeguard the environment are credible and given equal weight to those pursuing economic or social objectives. In short, a 'three legged stool' depends on each of its legs to stay upright. Since 1997, however, poor policy carpentry has made it look increasingly likely that Labour's stool will tip over.

The Beginnings of Retreat?

Although an 'unprecedented breadth of support for a radical strategy' was reflected in the energy of the government's early words and actions, its policy rhetoric was already noticeably softer by the time of the publication of *A New Deal for Transport*, the first transport White Paper in 20 years.[42] Although it was clearly less radical than it might have been (particularly in light of the hopes raised by the government's earlier rhetoric), the White Paper was still greeted with considerable enthusiasm. In particular, it set out a very reasonable analysis of the range of transport-related problems to be tackled, including road traffic growth and congestion, local air quality, social exclusion, climate change, urban sprawl, and rural sustainability, but with its relatively modest policy measures, it fell 'short of the promised radicalism and vision', and some critics even considered it to be a 'poorly focused and indecisive document'.[43]

In many ways, the White Paper can be seen as the beginning of Labour's nervousness over the possible political reaction to radical transport policies. Potentially significant interventions, such as motorway tolling and retail car parking charges, were dropped from the final document at the last minute, following media discontent and concerted lobbying from particular business groups such as the major supermarkets. The language had also changed – rather than an explicit focus on 'sustainability', the document praised the virtues of 'integrated transport', and even revisited the rhetoric of 'choice' which had underpinned the Conservatives' championing of roads-based policies a decade earlier.

This shift in rhetoric towards the notion of 'integrated' rather than 'sustainable' transport is important, especially since the idea of what exactly 'integration' meant was never really made clear. Was it improved physical integration between buses and trains to make public transport more attractive? Was it integration between the car and public transport through policies such as park and ride? Or was it a more general integration between policies designed to improve the transport system in other ways? Such uncertainty perhaps illustrates the oft-claimed divisions between John Prescott and Tony Blair on transport policy. Despite Prescott's apparent enthusiasm for making a genuine attempt to follow the sustainability agenda through actually reducing car use, by the time the White Paper was eventually released a year after Labour came to power, the Prime Minister's desire for less radical, more business- and voter-friendly policies based on ephemeral ideas such as 'integration' and the 'Third Way' between market provision and state regulation, was in the ascendancy.

In pursuing the mantras of 'integration' and 'choice', the White Paper had much more to say about potential 'carrots' designed to entice motorists out of their cars, rather than the more powerful 'sticks' fashioned to force them out. Carrots included proposed improvements to public transport (increased service frequencies, extended hours of operation, higher quality vehicles, enhanced integration, accessible real-time information), support for personal modes (cycling, walking), integrated land-use policy and attempts to influence an overall change in attitude.

In contrast, the 'sticks' were either not taken up, or were only addressed indirectly. True, policies such as bus priority or congestion pricing measures are not always physically or politically easy to enact.[44] But they have been reasonably well documented as successful within Europe, and even quite radical policies – such as the comprehensive urban road pricing systems introduced in the Norwegian cities of Oslo, Bergen and Trondheim – have found favour among initially sceptical electorates.[45] The challenge for Labour was to demonstrate that there was substance behind its words on sustainability by articulating the benefits of these and similar policies in terms of reduced congestion, better environmental quality and improved public transport.

Overall, *A New Deal for Transport* gave the distinct impression that, despite its words, the government was not wholly committed to tackling the root of the transport problem, that is the unsustainability of current transport patterns caused by car dependence as opposed to simply car ownership and responsible use. The result of the May 1997 general election was, at least partly, regarded as a reflection of majority support for a government which would lead and inspire public opinion through taking a principled stand on 'hard choices' such as the negative externalities of car dependence. Yet radical measures to reduce the impact of the car were quickly assumed to be electorally unpopular, because they affect the politically crucial sections of society who have become the most sub- or ex-urbanized, and hence car dependent. Much of what 'middle England' consumes – exclusive suburban estates, extensive convenience shopping – results in over-use of the car and the corollary of continued inner urban decay.[46] Intervening to address these trends in the name of sustainability would require a fundamental change in the lifestyle of the 'Mondeo Man' that brought Labour back to power, and it was this realization that forced the government's radical transport policies into reverse.

From Radicalism to Pragmatism

Two years after the publication of the White Paper, Labour formalized its vision of what it could achieve in *Transport 2010: The 10-Year Plan for Transport*.[47] The *10-Year Plan*'s headline figure of £180 billion in 2000 cash terms (£150 billion in real terms) of investment over ten years was broadly welcomed as representing a significant increase in transport spending, which would begin to close the gap in transport spending between the UK and its major European competitors (Figure 1.3). However, closer examination revealed a considerable degree of uncertainty over many of its forecasting assumptions, and over whether the planned resources were likely to materialize in the later years of the plan.[48] Of particular concern is whether the proposed split between public and private finance will be achievable in practice. As the House of Commons' Select Committee on Transport, Local Government and the Regions noted in its review of the *Plan* in 2002:

> The Plan must not be undermined by funding uncertainty. There are concerns, particularly for the railways, that the necessary levels of private sector support may not be forthcoming either at the right time or on the right terms . . . A more detailed breakdown of future expenditure for all aspects of the Plan is required if it is indeed to be a Plan rather than a wish list.[49]

With long-term resource allocation inevitably subject to significant uncertainty, the most important aspect of the *10-Year Plan* was the way in which it confirmed a change in the government's aspirations for sustainable

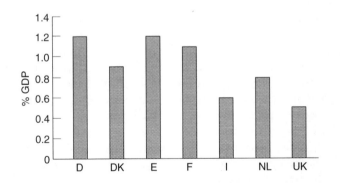

Figure 1.3   Transport infrastructure investment as a share of GDP, 1995. Source: Commission for Integrated Transport (2000) *European Best Practice in Delivering Integrated Transport – Key Findings*. CfIT, London.
   Key: D = Germany, DK = Denmark, E = Spain, F = France, I = Italy, NL = Netherlands, UK = United Kingdom.

transport policies. One of the *Plan*'s most striking features is the return to identifying congestion as the most important transport problem that policy and investment must overcome. Environmental problems, which are generally summarized as local 'pollution' rather than the emissions that threaten global sustainability, follow in second place. As Phil Goodwin – formerly the government's most senior independent transport advisor – notes in the final chapter of this book, the formal setting out of priorities at odds with previously articulated policy strategies suggests that the publication of the *10-Year Plan* marks the point when the government's retreat from a more sustainable policy agenda was made real.

Targeting of congestion as the primary problem affecting the transport system allowed ministers to sidestep the rather more difficult pursuit of a real reduction in the overall level of traffic. Such a change of priority revealed that, as for the Conservatives previously, the potential for transport investment to address *short-term* economic imperatives, rather than longer term objectives such as safeguarding the environment, lies at the heart of Labour's transport policy. Just as the 'new realism' – a normative policy position proposed by Goodwin which challenges continued large-scale road building – reflected the economic impossibility of predict and provide as much as it did emerging environmental concerns, so the focus on 'solving congestion' rather than reducing overall levels of road traffic was inevitable given the ingrained culture of car dependence and parlous state of the public transport system in the UK (Chapter 10).[50]

In its early months, the government had looked towards road pricing as the most credible solution to the congestion problem. Many transport academics and professionals, led by Goodwin, argued that the price mechanism could be applied to ration the supply of road space in the same way as any other scarce resource. The elegance of road pricing as a remedy for congestion is that it generates large revenue streams for investment in quality public transport alternatives, as well as prioritising scarce road space towards high value users. This in theory encourages further modal shift away from the car, improving economic efficiency by reducing congestion, and at the same time helps promote social inclusion by providing better transport options for disadvantaged groups with low levels of car ownership. Despite claims to the contrary, road pricing can also be an egalitarian, redistributive policy, since the costs of congestion are more clearly attributed to those who cause it, rather than being indiscriminately ascribed through general taxation.

Yet in the face of increasing public protests against the high cost of motoring and other policies perceived as being 'anti-car', the government quickly became reluctant to use pricing as a direct instrument for reducing traffic levels. Although support for some road pricing in the form of urban congestion charging in was maintained long enough for enabling legislation to be included in the Transport Act (2000), the government delayed its likely introduction by handing responsibility for implementing charging schemes to the local level.[51] The Mayor of London, Ken Livingstone, has used his unique personal mandate to pioneer the implementation of a congestion charging scheme for inner London, but it is uncertain that many provincial cities will quickly develop similar schemes. This is because the potential risks of political unpopularity and economic competition from other centres that resist charging, either through road pricing or by other means such as nonresidential parking levies, are just too great for most local politicians seeking re-election (Chapter 4).

The cost of this abandonment of demand management through pricing at the national level is a return to a policy of boosting the supply of mobility through increased road and rail capacity. Ironically, this means the government is faced with a policy paradox of its own making – by rejecting pricing, not only is congestion not directly suppressed, but potential revenue streams for new infrastructure are ruled out. This means the government will need to work (even) harder at expanding infrastructure capacity, but with greater constraints on available resources than would otherwise be the case. This is bad enough, but the need to meet self-imposed congestion targets – these were included in the *10-Year Plan* (and are discussed further in Chapter 10) – with restricted budgets also exerts considerable geographical bias on the government's strategic priorities. For example, the need to tackle traffic in the massively congested South East of England logically dictates that the majority of investment in the National Rail network should be directed to that region, even if this is at the risk of 'improvements north of Watford . . . [being] put on hold or axed'.[52]

The government's second policy shift concerns its strategy for minimising the environmental impact of the car. What is particularly revealing is the way in which it chose to interpret the evidence and advice given to it, which has had the effect of shifting the transport policy debate away from some of the 'hard choices' required to pursue a radical sustainable agenda. The government's attitude to the work of the Commission for Integrated Transport (CfIT), which played a major role in assimilating the knowledge on which the strategies of the *10-Year Plan* were based, is a particular case in point. Early CfIT advice to government was positive that genuine road traffic reduction, especially in urban areas, was attainable: 'Over time it should be possible to reduce traffic in the areas where most people live; we recommend that the Government should work in this direction'.[53]

The government chose to quote CfIT's advice in a selective way, however. The same report, *National Road Traffic Targets*, was also used to underpin another, perhaps more fundamental, change in the government's attitude towards the environmental impact of road traffic. This is simply that ministers pinned their hopes on improvements in road vehicle technology (the so-called 'technological fix'), rather than reductions in the absolute level of road traffic, to play the major part in reducing carbon dioxide emissions beyond the level required by its Kyoto Agreement commitments. By 2010, it is forecast that incremental design improvements in conventional vehicle engines will account for a cut in carbon dioxide emissions more than twice as large as that attributable to modal shift resulting from the package of public transport improvements contained in the *10-Year Plan*.[54] In the longer term, it seems that the government is hoping that the adoption of new forms of motive power, such as 'ecocars' running on hydrogen fuel cells, might prevent the renewed increases in emissions forecast in the *10-Year Plan* if the substantial technical problems can be overcome.[55] CfIT has also suggested that the level of 'transport intensity' might decrease, with future economic growth less dependent on increased mobility as previously. The evidence for this is mixed, however, with some studies reporting an increase in transport intensity during the 1990s.[56]

CfIT also claimed that even with modest policy intervention to reduce the *rate* of car traffic growth (that is, assuming continued increases in actual traffic levels), 'far more substantial reductions (up to 75 per cent) are forecast in the [nitrous oxide] emissions and [particulate] emissions that affect local air quality'.[57] This perhaps explains the prominence given by the government to local air quality targets, since if these forecasts are accurate, ministers can be confident that substantial environmental improvements can be achieved with little or no need to introduce policies aimed at actually curbing car use.

Taken together, these shifts in priority suggest that the aspiration to reduce the need to travel seems almost to have disappeared from Labour's agenda. In the middle of the 'environmental turn' of the 1990s, the Major government's *Planning Policy Guidance Note 13: Transport* (PPG13) clearly stated that in future, 'plans should aim to reduce the need to travel, especially by car'.[58] Management of the demand for travel was to become a favoured policy strategy, with direct intervention to reduce car traffic, most notably through the implementation of the fuel tax escalator as recommended by the RCEP. But it now seems that Labour is replacing this with the much more laissez-faire approach of giving people even more choice of whether and how to travel, implying that public transport will have to continue to compete for trips in a market system, where many car journeys remain underpriced in terms of their true economic, environmental and social costs.[59] This also chimes with public opinion. Research commissioned by CfIT during the development of the *Plan* demonstrated that reducing congestion in towns and cities was the top priority for the public after improved road maintenance, with reducing congestion on motorways and other major roads also figuring strongly.[60] In other words, a substantial constituency for more sustainable transport policies has emerged, as increasing numbers of people felt their quality of life to be significantly diminished by the impacts of road traffic. If ever there was an opportunity for a government to demonstrate leadership by seeking to convince the public that radical initiatives such as pricing were necessary to address their underlying concerns, this was it.

It could be argued that one of the most fundamental underlying messages in the *10-Year Plan* – indeed the message which marked the *Plan* out as the point where Labour's retreat from sustainable transport became fully apparent – was the much reduced emphasis placed on demand management. Although it offered the scenario that up to 20 towns and cities will have introduced some form of charging by 2010, this is likely to prove a hugely optimistic assumption, which neatly disguises the political choice made to omit other potential demand management measures such as motorway tolling. Instead, given that there was virtually no emphasis on reducing the need to travel, the only way open to the government to reduce congestion was to rely on a strategy of investing heavily in new and expanded transport infrastructure to accommodate as much mobility as possible, albeit by tweaking modal split to enhance the role of the railways. As a result, significant road building was back on the agenda.

The government's roads policy has now become quite adventurous, with the *10-Year Plan* containing among its commitments a very significant element of new road building. Jon Shaw and William Walton's study of the *Plan*'s trunk-road building proposals for England found that the mean number of roads completed each year from 2008 could reach 35, and so exceed that achieved by the Conservatives at any stage during their 1979–97 administrations (Chapter 4).[61] Shaw and Walton characterize Labour's approach as 'pragmatic multi-modalism', where high(er) levels of road building are pursued alongside enhanced public transport investment to produce a policy compromise based on what is politically realistic to deliver.

A similar picture has emerged in Scotland, where the devolved Scottish Executive has followed a transport policy path very similar to that of the UK government. After initiating a wide-ranging review of inherited roads schemes which led to the shelving of many long-standing projects, Scottish ministers floated radical demand management measures, including motorway tolling and workplace parking levies, just two weeks after the establishment of the devolved institutions in 1999.[62] But, as in England, by the time early words had crystallized in the form of the *Transport Delivery Plan*, Scotland's transport policy was again dominated by major road building without any central government commitment to new transport charges.[63] In Wales, devolution also pushed policies in the direction of improved internal road links, since these had been neglected at the expense of connections to England in previous years. Across Great Britain, therefore, the contrast from the early optimistic days of 1997 could not be more apparent: whereas the government then saw 'new roads as a last resort rather than a first', today its commitment to more sustainable strategies is much less clear.[64]

Back to the Car-Owning Democracy?

With Labour's commitment to the sustainable transport agenda already in doubt for many commentators, the remarkable events of September 2000 demonstrated both that the car-owning democracy was very much in evidence, and that the government was willing to appease motorists' demands. For almost two weeks, Britain witnessed unprecedented direct action as farmers blocked fuel depots and truckers created 'go-slow' convoys choking the motorway network, with petrol stations running dry as a result. As the economy and essential public services teetered on the verge of collapse, the government faced a defining moment in the development of its transport policy: whether to capitulate to the protestors' demands for an immediate significant reduction in fuel taxes, or to keep faith with the policy of steadily increasing fuel taxes in order to restrain the growth of traffic.

The seriousness of the government's *volte face* on transport taxation and, by implication, its attitude to sustainable transport more generally, was brought into sharp focus by its response to the fuel tax protests. Having already abolished the fuel tax escalator early in 2000 just a matter of weeks after publishing its *Climate Change Strategy*, the government found itself trailing in the opinion polls for the first (and only) time in the 1997–2001 parliament.[65] Desperate to reverse the situation with only months to go before its preferred date for the general election, the Chancellor announced a two pence per litre cut in fuel duty in his November Pre-Budget Report, along with further reductions for lower-sulphur fuels, claiming that the rate of fuel tax had 'no impact on traffic levels', and was 'not designed to do so'.[66] These actions were backed up with statements from the Prime Minister, who ad hoc abandoned the earlier policy that increases in fuel tax revenues would be hypothecated to public transport schemes, stating that the treatment of money from fuel taxes was to remain as any other part of the general revenue stream.[67] Throughout the election campaign that followed, Blair repeated the twin mantras of 'investment' and 'choice' in outlining his strategy for transport, adopting a tone completely at odds with his previous statements promoting a 'coalition for the environment'.[68]

In essence, these decisions made it clear that the government had chosen to abandon its previous strategy of articulating the environmental case for higher fuel taxes as pioneered by the RCEP, in favour of populist cuts in taxation at the altar of political expediency. And this at a time when evidence was emerging that showed British motorists were not nearly as badly off as the protestors liked to make out. First, the real costs of running a car had remained stable for 25 years while rail and bus fares had risen by 50 per cent and 75 per cent respectively (Figure 1.4).[69] Second, CfIT demonstrated that when the *total* level of car taxation is taken into account rather than focusing on fuel costs, UK motorists were *not* particularly highly taxed compared to others in Europe (Table 1.1).[70] Third, calculations suggested that there remained a significant gap – up to £24 billion in 1998 – between the amount of revenue raised through motoring taxes and the overall cost to society of road vehicles.[71] Finally, and perhaps most damning for the government, was that evidence suggested that the increases in fuel tax might be just beginning to work. Writing days before the pre-election Budget of 2001, David Begg urged the government to resist the pressures to cut fuel taxes since:

(in 2000) road traffic in the United Kingdom grew by only 0.3 per cent – one of the lowest increases ever recorded in the modern age. What makes this volume all the more surprising is that it coincided with an economic boom: GDP was growing by 3 per cent. Between 1998 and the final abolition of the fuel tax escalator in 2000, the rate of traffic growth stabilized for the first time under conditions of economic growth.[72]

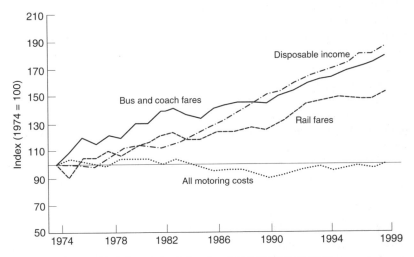

Note: 'All motoring costs' includes petrol and oil costs and cost of vehicle purchase

Figure 1.4   Real changes in the costs of transport and disposable income, 1974–99. Source: Department of the Environment, Transport and the Regions (2000) *Transport 2010: The 10-Year Plan*. DETR, London.

Table 1.1 Comparison of total taxes on car ownership and use across Europe for 2000, £ sterling at purchasing power parity, ranked for 1600cc engine size

		Engine Size		
Rank	Country	1,000cc	1,600cc	2,000cc
1	Netherlands	978	1,295	2,096
2	Finland	800	1,032	1,565
3	Denmark	723	1,024	1,551
4	Ireland	652	1,006	1,467
5	UK	731	976	1,201
6	Italy	758	968	1,301
7	France	756	955	1,191
8	Belgium	627	906	1,233
9	Greece	548	823	1,581
10	Norway	644	809	1,217
11	Germany	565	747	962
12	Sweden	582	743	982
13	Spain	546	709	1,035
14	Luxembourg	392	524	691

Source: Commission for Integrated Transport (2001) *European comparison of taxes on car ownership and use.* CfIT, London.

But perhaps the most striking illustration of the irony of the fuel tax protests was to be drawn from the evidence that later emerged of the significant changes in transport behaviour they brought about. CfIT reported very significant reductions in traffic levels of up to 39 per cent on motorways and 25 per cent on other major roads, with people making much more informed choices about when to use their cars, and for which journeys.[73] Around 75 per cent of people changed their travel behaviour in some way, with 20 per cent of parents abandoning the 'school run' in favour of walking, cycling or other modes.[74] Many train companies reported increases in patronage of up to one quarter, demonstrating that significant modal shift away from the car could be achieved in certain circumstances without heavy investment in new infrastructure. Finally, huge improvements in urban air quality were measured across Britain's major cities.[75]

The extent to which this about-turn in the government's transport strategy resulted directly from public resistance to its earlier promotion of so-called 'anti-car' policies such as road pricing, the fuel tax escalator and reducing the roads programme is contested. Writing in the summer of 1999, Phil Goodwin concluded that this 'backlash' was a temporary phenomenon caused by a lack of perceived improvements in public transport despite road traffic reduction measures and increased taxes on car ownership and use.[76] But this was before the fuel tax protests demonstrated just how powerful the car owning democracy remained, particularly for a Labour government dependent on the support of swing voters in 'middle England' in its bid for a second (and now, third) term.

It has become clear that, despite the rhetoric of its early days, the Labour government's record in office marks something of a return to policies of the 1980s.[77] Most ominously for advocates of the sustainable transport agenda is the surprisingly large proportion of promised resources directed towards roads projects.[78] Put simply, the evidence outlined in this chapter demonstrates that the government seems to have rejected the core objective of sustainability, that of managing the demand for transport. Instead, partly because of its focus on growing the economy, partly because of the promise of the 'technological fix' as an escape route from the worst of the environmental impacts of car traffic, and partly because of the public backlash against transport policies perceived as anti-motorist, Labour has reverted to a transport strategy designed to accommodate much greater mobility, albeit with parallel investment in the railways and local public transport in an attempt to tweak overall modal share. If a real, sustained increase in public transport investment can be delivered over the government's second term, this will represent a welcome and genuine break from the past, since increased resources for roads have invariably been found at the expense of public transport. But there remain substantial doubts about whether this will really happen, especially given the concerns about the level of commitment that both the government and the private sector have in seeing the level of funding envisaged in the *10-Year Plan* actually delivered.

Whether Labour's retreat on sustainable transport represents a pragmatic response to the dilemma of how to deal with the renewed mobility demands of a steadily growing economy, a more cynical capitulation to vocal demands for an end to 'anti-car' policies, or a combination of the two, is a matter for debate. The House of Commons' Transport Select Committee for one was in no doubt when it decided that the government was 'mistaken' and 'wrong' in its overall transport strategy, that it formulated policy statements that were 'ill balanced', 'incomprehensible' and 'over-optimistic,' and that it relied too much on 'casual enthusiasts' to reassure it that it is delivering on its promises.[79] But what is certain is that the government's policy stance has changed significantly over its first six years. Although Labour's transport policy was originally founded on 'choice', this choice was aimed at reducing car dependence to tackle car-related congestion and pollution. What the policy strategies and investment programme embodied in the *10-Year Plan* are likely to result in is a different kind of choice: choice to travel to more places, by more modes, more often.[80] There are many carrots to promote all kinds of travel, but very few sticks to prevent unnecessary mobility. Government rhetoric envisions a less congested, more reliable future for transport, but ministers are unwilling to back up their words with radical policies to alter the structures of transport governance and argue the case for sustainable transport policies so that this future can be created. It seems that the car-owning democracy is alive and well.

Introducing the Rest of the Book

The remainder of the book analyses the government's transport strategies and record of delivery in greater detail. The first section, 'Policy and Politics', continues by examining the roles of the different levels of government in the UK in implementing the sustainable transport agenda. In Chapter 2, Austin Smyth reviews how devolution, one of Labour's 'big ideas', has influenced the development of transport policies by different administrations across the United Kingdom. The theme of central-local relations and their impact on transport policy is developed further in Chapter 3, where Geoff Vigar and Dominic Stead examine the role of local authorities in delivering Labour's transport agenda.

The central section of the book, 'Progress in Policy Implementation', assesses Labour's record with respect to each of the UK's main transport modes. In Chapter 4, William Walton critiques Labour's road-building policy since 1997, then in Chapter 5, Jon Shaw and John Farrington question whether the promised 'railway renaissance' is likely to materialize. Chapter 6 is also concerned with rail transport, but focuses specifically on light rail systems and the London Underground. Richard Knowles and Peter White analyse the 'stop-go' story of investment in light rail and the 'Tube', and ask what Labour's approach to big city transport reveals about its commitment to the sustainability agenda more widely. The government's recognition of the vital contribution the bus industry could make to its transport plans is scrutinized by John Preston in Chapter 7, and Rodney Tolley then appraises Labour's policy on the so-called 'personal' or 'benign' modes – walking and cycling – in Chapter 8. In the last chapter of this section, Brian Graham addresses the expansion of air travel, one of the most profound challenges facing any government in its desire for better sustainability.

Looking towards 'The Future', the book's final chapter assesses the prospects for the implementation of genuinely sustainable transport policies over the next five years and beyond. Written by Phil Goodwin, formerly the Labour Government's most senior independent transport advisor, the chapter develops the issues raised in the book to construct a critical overview of Labour's performance in terms of its own policy goals. Although it was always likely to be difficult to achieve a genuine shift in UK transport policy towards even politically realistic sustainable outcomes –

Goodwin argues Labour has found itself in a position that would have confronted any government – the chapter offers a view on how a credible sustainable transport agenda might develop. Applying his experience at the heart of transport policy decision making, Goodwin outlines how attitudes in the UK will have to change if tangible steps towards a more sustainable agenda for transport are to be made.

NOTES

1 Labour Party (1996) *Consensus for change: Labour's transport strategy for the 21st century.* Labour Party, London.
2 Pucher, J and Lefèvre, C (1996) *The urban transport crisis in Europe and North America.* Macmillan, London.
3 Department of the Environment, Transport and the Regions (1998) *A new deal for transport: better for everyone.* Cmnd 3950, The Stationery Office, London, 3.
4 Department of the Environment, Transport and the Regions (1998) *A new deal for transport.*
5 Daniels, P and Warnes, A (1980) *Movement in cities.* Methuen, London, 4.
6 Schaeffer, K and Sclar, E (1975) *Access for all: transportation and urban growth.* Penguin, London.
7 See, for example, Meyer, J and Gomez-Ibanez, J (1981) *Autos, transit and cities.* Harvard University Press, Cambridge, MA.
8 Wistrich, E (1983) *The politics of transport.* Longman, London.
9 Wright, F (1963) A verbatim record of a symposium held at the School of Architecture from March to May 1961. In *Four great makers of modern architecture. Gropius, Le Corbusier, Mies van der Rohe, Wright.* Columbia University School of Architecture, New York, 8.
10 Webb, S and Webb, B (1963) *The story of the King's highway.* New edition. Cass, London, 147.
11 Abercrombie, P (1945) *Greater London Plan 1944.* HMSO, London; Abercrombie, P (1949) *The Clyde Valley Regional Plan 1946.* HMSO, Edinburgh.
12 Starkie, D (1972) *The motorway age.* Pergamon, Oxford.
13 Ministry of Transport (1963) *Traffic in towns.* HMSO, London.
14 Goodwin, P (1999) Transformation of transport policy in Great Britain. *Transportation Research Part A*, 33, 657.
15 Westwell, A (1991) *Public transport policy in conurbations in Britain.* Unpublished PhD Thesis, Keele University, Keele.
16 For a discussion on car dependence, see RAC (1995) *Car dependence.* RAC Foundation for Motoring and the Environment, London.
17 BBC (1999) *Transport policy: what you want.* http://news.bbc.co.uk/1/hi/uk/294394.stm (accessed 15 December 2002).
18 Schaeffer, K and Sclar, E (1975) *Access for all*, 103–4.
19 Roberts, J; Clearly, J; Hamilton, K and Hanna, J (eds) (1992) *Travel sickness.* Lawrence and Wishart, London.

20   Goodwin (1999) Transformation of transport policy in Great Britain, 658.

21   Department of Transport (1989) *Roads for prosperity*. HMSO, London.

22   Organisation for Economic Development and Cooperation (1979) *Report of the seminar on urban transport and the environment*. OECD, Paris, 149.

23   United Nations World Commission on Environment and Development (1987) *Our common inheritance*. Oxford University Press, Oxford.

24   Goodwin, P (1999) Transformation of transport policy in Great Britain, 661.

25   Black, W (1998) Sustainability of transport. In Hoyle, B and Knowles, R (eds) *Modern transport geography*. Second, revised edition. Wiley, Chichester, 337–51. A further definition of sustainability in transport is offered by Greene and Wegener, based on Daly (1991): '(i) its [transport's] rates of use of renewable resources do not exceed their rates of regeneration; (ii) its rates of use of non renewable resources do not exceed the rate at which sustainable renewable substitutes are developed; and (iii) its rates of pollution emission do not exceed the assimilative capacity of the environment.' Greene, D and Wegener, M (1997) Sustainable transport. *Journal of Transport Geography*, 5, 177–90. This definition is referred to in Chapter 9.

26   At the risk of undermining the main argument of this volume (!), there are, of course, those in the scientific community who maintain that climate change is not happening, not relevant to the sustainability of the environment and/or not caused by human action.

27   Royal Commission on Environmental Pollution (1994) *Eighteenth report. Transport and the environment*. HMSO, London.

28   Goodwin, P; Hallett, S; Kenny, P and Stokes, G (1991) *Transport: the new realism*. Transport Studies Unit, University of Oxford.

29   Shaw, J and Walton, W (2001) Labour's new trunk-roads policy for England: an emerging *pragmatic multi-modalism. Environment & Planning A* 33, 1131–1156.

30   Labour Party (1996) *Consensus for change*.

31   *Hansard* (1998) Volume 309, 25 March, 468.

32   John Prescott, quoted in *The Guardian* (1997) 6 June.

33   Shaw, J; Walton, W and Farrington, J (2003) Assessing the potential for a 'railway renaissance' in Great Britain. *Geoforum*, 34, 141–56; Docherty, I (2002) TGRG Page. *Journal of Transport Geography*, 10, 319–20.

34   Department of the Environment, Transport and the Regions (1997) *New roads as a last resort*. Press Release 216, DETR, London.

35   Glaister, S (2001) *UK transport policy 1997–2001*. Address to the British Association for the Advancement of Science, Glasgow, September, 1.

36   Both the methodology used to calculate this figure, and the figure itself, are contested, but there nevertheless appears to be consensus that the economic cost of congestion is highly significant.

37   See Kresl, P (1995) The determinants of urban competitiveness: a survey. In Kresl, P and Gappert, G (eds) *North American cities and the global economy*. Sage, Thousand Oaks, CA, 45–68.

38   Hamer, M (1996) Clean air strategy fails to tackle traffic. *New Scientist*, 6 August.

39 Department of the Environment, Transport and the Regions (1998) *A new deal for transport*, 26.

40 Reed, M (2001) *Strathclyde Passenger Transport – its plans for the future in the light of the Transport (Scotland) Act*. Address to Moving Scotland Forward conference, Edinburgh, 5 February.

41 Ophuls, W (1977) *Ecology and the politics of scarcity*. WH Freeman, San Francisco.

42 Goodwin, P (2001) *The nine year plan for transport: what next?* Paper presented to the Transport Planning Society, London, July, 11.

43 Docherty, I and Hall, D (1999) Which travel choices for Scotland? A response to the government's White Paper on integrated transport in Scotland. *Scottish Geographical Journal*, 115, 193–210; Glaister, S (2001) *UK transport policy 1997–2001*, 3.

44 Johansson, B and Mattson, L-G (eds) (1995) *Road pricing: theory, empirical assessment and policy*. Kluwer, Amsterdam.

45 Odeck, J and Brathen, S (1997) On public attitudes towards implementation of toll roads – the case of the Oslo Toll-Ring. *Transport Policy*, 4, 77–83; Larsen, O (1995) The toll cordons in Norway. *Journal of Transport Geography*, 3, 187–97.

46 The phrase 'middle England' seeks to encapsulate the generally affluent, often suburban voters on whom every government depends for its parliamentary majority. It is one of the ongoing ironies of transport policy that these are also the people who, because of their lifestyles, are usually the most car dependent, and thus most sensitive to tax increases or other policies that could be perceived as 'anti-car'.

47 Department of the Environment, Transport and the Regions (2000) *Transport 2010: The 10-Year plan for transport*. DETR, London. It is important to note that (especially) following devolution, central government policy documents such as the *10-Year Plan* have different applicability across the UK. In this case, for example, the *10-Year Plan* was relevant to: roads only in England, since responsibilities for roads elsewhere were devolved to the national assemblies/parliament; and rail in Great Britain but with much of the detail obscured since responsibility for railways in Scotland was *partly* devolved to the Scottish Parliament.

48 See Goodwin, P (2001) *The nine year plan for transport* and Glaister, S (2000) *Transport policy, control and value for money*. Imperial College, London, for detailed critiques of the methodologies and assumptions of the *10-Year Plan*.

49 House of Commons (2002) Session 2001–2002, HC 558–I, 27 May, 133. http://www.publications.parliament.uk/pa/cm200102/cmselect/cmtlgr/558/55802. htm (accessed 15 December, 2002).

50 See Goodwin, P et al. (1991) *Transport: the new realism*, and Goodwin, P (1997) *Solving congestion*. Inaugural lecture for the professorship of transport policy, University College London.

51 Transport Act (2000) *Public general Acts – Elizabeth II*. Chapter 38. The Stationery Office, London.

52 *Financial Times* (2001) 11 October, 10.

53   Commission for Integrated Transport (1999) *National road traffic targets*. CfIT, London, 19.

54   Department of the Environment, Transport and the Regions (2000) *Climate change: the UK programme*. Cmnd 4913, The Stationery Office, London.

55   Bannister, D (2000) Sustainable urban development and transport – a Eurovision for 2020. *Transport Reviews*, 20, 113–130.

56   Standing Advisory Committee on Trunk Road Assessment (1998) *Transport investment, transport intensity and economic growth: interim report*. The Stationery Office, London.

57   Commission for Integrated Transport (1999) *National road traffic targets*, 3.

58   Department of Transport and Department of the Environment (1994) *Planning policy guidance note 13: transport*. DoT/DoE, London, 3.

59   Royal Commission on Environmental Pollution (1994) *Eighteenth report*.

60   Commission for Integrated Transport (2000) *Public attitudes to transport in England*. CfIT, London.

61   Shaw, J and Walton, W (2001) Labour's new trunk-roads policy for England.

62   Scottish Executive (1999) *Travel choices for Scotland: strategic roads review*. http://www.scotland.gov.uk/travelchoices/docs/tcfs-00.htm; Scottish Executive (1999) *Tackling Congestion*. The Stationery Office, Edinburgh. http://www. scotland. gov.uk/library2/doc01/taco-00.htm (both accessed 15 December, 2002).

63   Scottish Executive (2002) *Scotland's transport: delivering improvements*. The Stationery Office, Edinburgh. http://www.scotland.gov.uk/library3/transport/ stdi-00.asp (accessed 12 December, 2002).

64   Department of the Environment, Transport and the Regions (1997) *National road traffic forecasts (Great Britain) 1997*. DETR, London.

65   Department of the Environment, Transport and the Regions (2001) *Climate change: the UK programme*. The Stationery Office, London. http://www.defra. gov.uk/environment/climatechange/cm4913/ (accessed 13 November, 2002).

66   The Chancellor of the Exchequer, Gordon Brown, made these claims on BBC Radio 4's *Today* programme in advance of his pre-Budget Report on 8 November, 2000.

67   BBC (2001) *Tony Blair quizzed*. http://news.bbc.co.uk/vote2001/hi/english/ forum/newsid_1216000/1216175.stm (accessed 15 December, 2002).

68   Blair, T (2000) A new coalition for the environment. *The Guardian*, October 24.

69   Department of the Environment, Transport and the Regions (2000) *Transport 2010*.

70   Commission for Integrated Transport (2001) *European comparison of taxes on car ownership and use*. CfIT, London.

71   OXERA (2000) *The wider impacts of rail and road investment*. The Railway Forum, London; Department of the Environment, Transport and the Regions (1998) *A new deal for transport*.

72   Begg, D (2001) Hit the brakes. *The Guardian*, 6 March.

73   Commission for Integrated Transport (2001) *Lessons of the September 2000 Fuel Crisis*. CfIT, London.

74   Lyons, G and Chatterjee, K (eds) (2002) *Transport lessons from the fuel tax protests of 2000*. Ashgate, Aldershot.

75   Seakins, P; Lansley, D; Hodgson, A; Huntley, N and Pope, F (2002) New directions: mobile laboratory reveals new issues in urban air quality. *Atmospheric Environment*, 36, 1247–1248.

76   Goodwin, P (1999) Transformation of transport policy in Great Britain.

77   Shaw, J and Walton, W (2001) Labour's new trunk-roads policy for England.

78   Many commentators (including the editors of this volume) recognize the continued need for *some* road building in particular places under particular circumstances. See Standing Advisory Committee on Trunk Road Assessment (1999) *Transport and the Economy*. The Stationery Office, London, for a discussion of where road building can be most effective.

79   House of Commons (2002) Session 2001–2002, HC 558–I, 27 May. The 'casual enthusiast' jibe was without doubt directed at Lord Birt, the former Director General of the BBC, whom the Prime Minister appointed as his Special Advisor on Transport despite his lack of experience in the field. Birt's proposals – which included the notion that a new network of tolled 'super-motorways' should be built across Britain either alongside or above existing routes – were the subject of some derision in the national press and professional transport community.

80   Walton, W and Shaw, J (2003) Applying the new appraisal approach to transport policy at the local level in the UK. *Journal of Transport Geography*, 11, 1–12.

2

Devolution and Sustainable Transport

Austin Smyth

Contrary to the common tendency of governments to centralize power so that they can exert maximum influence on policy formulation and delivery, Labour has fulfilled its long-standing commitments to introduce devolution to the United Kingdom.[1] Long regarded as one of the most centralized states in the developed world, the UK's political structure was substantively decentralized by devolution in 1999, with power over many domestic policy areas transferred from Westminster to the elected Scottish Parliament, National Assembly for Wales and Northern Ireland Assembly. Almost certainly, this transformation of the political landscape will prove irrevocable, and may herald the beginnings of a longer process of enhanced regional autonomy across the UK.

The attraction of devolution to its supporters is that it offers the potential to create local solutions to suit local circumstances. This creative potential has to be balanced, however, with the risk to the state that its devolved regions or countries might pursue policies that contradict central priorities. After all, this is what 'regional' or 'national' autonomy is all about. This chapter examines the experience of devolution to date, and analyses how the pursuit of a sustainable transport policy across the UK has been influenced by the actions of the new devolved institutions.

Despite the potential for policy inconsistency and incoherence – the physical and socioeconomic differences between England and Scotland, Wales and Northern Ireland are many and varied – it so happens that the UK has, to date, maintained a single transport strategy of sorts. The elements of this broad strategy are, in the main, common not just to the approaches of Westminster and the devolved institutions in Scotland, Wales and Northern Ireland, but also to those of the Greater London Assembly and English local authorities (Chapter 3). At the rhetorical level, they include a renewed emphasis on public transport with increased investment in buses and trains, initiatives to support walking and cycling, a transfer of freight from road to rail, and a determination to relieve congestion, reduce pollution and cutback on the emission of greenhouse gases. But it is not just the broad aims that are consistent, since the range of actions being considered or undertaken by the different bodies is also similar. Although the devolved administrations' interpretation of Labour's sustainable transport agenda varies according to local circumstances, there is nevertheless evidence of significant retreat from the aspiration to improve transport sustainability in each of the jurisdictions (Chapter 1).

Looking at the sequence of events, it might even appear as if devolution never really happened, and that Westminster dictated a common policy all along. That would be a misinterpretation. Although it is true that overarching institutions designed to afford a degree of coordination of government policies throughout the UK were established, they have not been used vigorously – at least in the transport policy arena – to ensure a common approach, since this would clearly be against the spirit of developing autonomy in the devolved nations. This is not to say, however, that mechanisms such as the Joint Ministerial Committee (JMC), a forum bringing together senior representatives of UK central government and the devolved administrations, might not prove valuable as a means to discuss issues of common interest.

Part of the reason for the alignment of transport policies was that the government's White Paper, *A New Deal for Transport: Better for Everyone*, was published in 1998 before the devolved assemblies were established.[2] But the underlying explanation for the continuing similarities in transport policy rhetoric, strategy and implementation lies in the nature of devolution itself. One of the paradoxes of the devolution process was highlighted as the new institutions assumed their powers, when the impacts on policy of the sheer novelty of the devolved assemblies quickly became apparent. The new politicians were faced with one particular dilemma shortly after their election – how was the popular support for change that created the new institutions to be expressed when policy makers were inexperienced in administration and thus naturally inclined towards incremental development of existing strategies rather than radical reappraisal?

For Scotland and Wales, the fact that Labour was in government at the devolved level (albeit in coalition) as well as at Westminster provided an additional disincentive to substantial shifts in policy that could be perceived as 'splits' in government. Put simply, if one administration changes policy under pressure, enormous strain is exerted on those other jurisdictions seeking to maintain their position. So, for example, when the Chancellor announced the abandonment of the fuel tax escalator, it instantly became much more difficult for any of the devolved administrations to promote sustainable transport (Chapter 1).

Uncertainty over the initial direction of the new institutions was reflected in public attitudes towards the early experience of devolution – the assemblies were widely perceived to have got off to a slow and shaky start as politicians accustomed themselves to their new roles. In Scotland, a poll claimed that 80 per cent of the population felt their Parliament had little or no impact. In Wales, where devolution had enjoyed weaker popular support, the *Western Mail* captured the public mood with the claim that the National Assembly had failed to make its mark: 'It is not yet strong; it has no real teeth'.[3] In contrast, the Northern Ireland Assembly is undoubtedly popular, as people seek political solutions to the ingrained social conflicts that have disfigured the province for decades. For many, its stop-start existence has been exasperating.[4]

A Spectrum of Devolved Powers

All of these experiences are related to the fact that, to date, devolution has been a piecemeal affair. The ad hoc nature of the whole project is reflected in the extent to which the devolved Scottish, Welsh and Northern Irish governments have varying responsibilities. Most importantly, whereas Scotland and Northern Ireland have been assigned primary legislative powers, the National Assembly for Wales has only secondary legislative powers. The Scotland Act (1998) provided the framework for the Scottish Parliament and its Members (MSPs). The Parliament has responsibility for key areas including health, education, economic development and local government as well as transport and the environment. In these matters, Holyrood (the Scottish equivalent of 'Westminster') is able to amend or repeal existing Acts of the UK Parliament and to pass new laws of its own. In Wales, where devolution was only narrowly approved in the 1997 referendum, the National Assembly in Cardiff has only limited powers to make secondary legislation. This allows Assembly Members (AMs) to make minor technical changes to some Acts passed at Westminster. Even then, the UK Parliament usually has the right to veto the resultant amendments. The Northern Ireland Assembly in Belfast has full legislative and executive powers over all devolved matters, which include health, education, economic development, culture and transport. The devolved Northern Irish Executive is drawn from Members of the Local Assembly (MLAs) representing the Province's two communities. In contrast with the rest of the UK, local councils in Northern Ireland have only very limited powers and as a consequence have little influence over transport policy (Chapter 3).

The devolution process in England has been even more hesitant. Conceived largely in response to the (in)famous 'West Lothian Question' – the situation where Scottish MPs at Westminster can vote on English domestic matters whilst English MPs have no equivalent competence over Scottish legislation – Labour finally committed itself to devolution for the English regions with the publication in May 2002 of the White Paper, *Your Region, Your Choice: Revitalising the English Regions*, which holds out the prospect of elected regional assemblies following consultation and referendums. Currently, very little power is devolved to the English regions, with regional government limited to the network of nine Government Offices for the Regions (GORs) responsible for overseeing the implementation of central government policies. More recently, Regional Development Agencies (RDAs) have been established across England as a first step towards political devolution. But since they are 'appointed by Ministers in London, they have none of the legitimacy that comes from election'.[5]

The most substantial devolution of power in England to date is that afforded to London. Shortly after the Scottish, Welsh and Northern Irish devolution legislation was passed, Londoners were given a measure of control over some of their own affairs with the creation of the Greater London Assembly, a 25 member chamber which scrutinizes the elected Mayor of London. In combination, the new London institutions are known as the Greater London Authority (GLA). Whilst not strictly 'devolution' – the Greater London Assembly has no law-making powers – the new arrangements for London give local politicians substantive powers over the development and implementation of public policy in the capital.[6] Amongst the Mayor's executive responsibilities are transport, the police, the fire service and economic development, and s/he controls four administrative bodies including Transport for London (TfL) (Chapter 6).

Control of transport policy: devolved or reserved?

The complexity of UK devolution means that it is not always clear where the locus of decision-making power for transport actually lies – each of the devolved jurisdictions has a different degree of control over its own transport system (Table 2.1). For example, the National Assembly for Wales has some influence over transport policy, but its activity is largely confined to the distribution of available resources since strategic direction remains largely determined by Westminster. Scotland and Northern Ireland have much more freedom of action. The Scottish Parliament has legislative control over most aspects of transport in Scotland, and passed its own Transport (Scotland) Act in 2001.[7] Although broadly similar in content to the UK Transport Act (2000), the Scottish legislation nevertheless contained a number of unique measures including a national concessionary fares scheme and enhanced provisions for the introduction of bus Quality Contracts compared to England (Chapter 7). In April 2001, Scotland also

Table 2.1 The transport sector policy framework: *de facto* devolved and reserved matters

	Scotland	Wales	Northern Ireland
Road	Totally	Limited	Totally
Rail	Partly	None	Totally
Bus	Totally	Limited	Totally
Air	Limited[1]	None[1]	None[1]
Sea (ferry)	Substantial	None	None

Note
1   It is possible for each of the devolved administrations to influence aviation policy through various secondary mechanisms such as the application of their planning powers (Chapter 9).

assumed responsibility for funding its domestic rail services and is setting out its own strategic position on rail policy. However, the Scottish Executive's influence over rail policy is very restricted since the Strategic Rail Authority (SRA) retains power over strategic rail development, investment prioritization and refranchising on a GB-wide basis.[8] In contrast, the Northern Ireland Assembly has complete control over roads, including motorways, buses and the limited rail network in the Province.

In London, the Mayor and Greater London Assembly have substantial transport powers. Through TfL, the GLA has responsibility for roads, buses, taxis, river services and strategic transport planning including new modes such as light rail, although the Underground has thus far remained under the control of the Secretary of State for Transport. Responsibility for the Tube has not yet been transferred to the GLA following a battle between ministers and the Mayor over its future financial structure (Chapter 6). In July 2001, the Mayor, Ken Livingstone, issued his first transport strategy, in which he identified a number of key priorities included reduced congestion, increasing investment in the Underground, and improving the quality and capacity of the bus and national rail networks in London.

It is easy, however, to overstate the GLA's powers to create an integrated transport plan for London. Policy control of the commuter railways remains with the SRA. This means that there is clear potential for conflict between the GLA and the SRA: the latter, with its national remit and limited budget, is likely to find it difficult to respond fully to the Mayor's and TfL's concerns about service quality, reliability and overcrowding on the London rail network. Although legislation appears to give the GLA significant influence over the SRA on policy in the London area, in practice, the GLA may only be able to do what the SRA will not. And, as Stephen Glaister has pointed out, the GLA can only adopt this degree of responsibility if it has the money to do so and, crucially, if the Secretary of

State does not prohibit it.[9] Indeed, Westminster's enduring veto over the Mayor's actions is reinforced by the fact that the s/he is required to ensure that Mayoral strategies (on transport, economic development and so on) are consistent with such national policy, including such international obligations, as the Secretary of State might bring to the Mayor's attention.[10]

Even where powers are in theory fully reserved to Westminster, the actual picture of which institution(s) are competent to act in particular areas of policy is not as clear as might be expected. Although some transport functions, such as vehicle licensing and testing, maritime safety and air traffic control, are entirely reserved – that is Westminster government departments retain responsibility in these fields across all of the UK – there are some important functions where the allocation of power is particularly complex. For example, the Scottish Executive has become increasingly involved in aviation, which is formally included in the list of reserved powers.[11] It has suggested using its planning powers to influence the UK's strategy for airport capacity, and its regional development powers to promote the designation of public service obligations and subsidies for selected marginal services in order to address problems of peripherality (Chapter 9). By contrast, ministers in Northern Ireland have so far felt unable to use their equivalent powers to become involved in air route development, since they perceive this entirely as a reserved matter.

The structure of taxation and finance for transport under devolution adds yet another layer of complexity to the policy making process. Under the current financial system, each of the devolved administrations receives an annual block grant from Westminster, which is determined on the basis of the so-called 'Barnett Formula', named after the former Chief Secretary to the Treasury, Lord Barnett, who established the current funding structure in 1978. The formula was introduced as a means of maintaining public spending in Scotland, Wales and Northern Ireland at a level designed to close the prosperity gap with England. It works by calculating total public spending under a range of 'domestic' or 'identifiable' policy headings – including health, education, social services, local government and transport – and allocating each of the UK's other constituent countries an enhanced share of the growth in the English total defined according to population and perceived need.

The generosity of the Barnett system is contested, particularly in Scotland where there is constant debate over whether the country is a net contributor or recipient of finance to/from the UK Treasury.[12] This uncertainty arises because the formula is not concerned with how much taxation revenue is raised in each jurisdiction, nor does it take into account 'nonidentifiable' budget headings – such as defence, macroeconomic transfers and research and development grants – which HM Treasury defines as UK 'national' spending, but which critics claim are disproportionately directed towards

England. But what is generally agreed is that the 'domestic' policy streams in the devolved nations are generously resourced in comparison to England. Evidence supplied by the Treasury to the House of Commons' Treasury Committee showed that in 1995/96, per capita spending on 'identifiable' headings in Scotland, Wales and Northern Ireland was respectively 32 per cent, 25 per cent and 32 per cent higher than in England.[13]

The importance of the continued existence of the Barnett Formula for transport policy is two-fold. First, it means that the devolved assemblies have a higher per capita financial allocation for domestic spending compared to England. Since the Scottish Parliament and Northern Ireland Assembly have a great deal of freedom in how they spend their block grant – they are not required to spend the same proportion of their resources on each heading as England does – they therefore have considerable discretion to prioritize (or deprioritize) transport matters. Second, this potential for substantially greater transport spending per head, coupled with the lower population densities of the devolved nations, tends to mean that transport policy is pushed in the direction of more road building.

A further complication arises from the fact that, since the Barnett Formula distributes only the *additional* expenditure over the historic baseline allocation, it is inevitable that the more generous per capita spending enjoyed by the devolved nations will be eroded over time. This means that Scotland, Wales and Northern Ireland could continue to lag behind England in terms of economic prosperity, yet see their share of the UK financial cake reduced significantly. It could be that this so-called 'Barnett Squeeze' might limit the amount of cash that would otherwise have been directed to transport infrastructure and services, unless the devolved administrations decided to reallocate spending from, say, health and education. With the devolved governments also committed to significant revenue spending on socially-necessary transport services, such as the islands ferries in Scotland, the level of funding available for transport *infrastructure* might come under even more pressure.

The final layer of complexity in devolution lies in the fact that central government retains enormous power to influence transport policy right across the UK because fiscal policy is reserved to Westminster. Since the devolved assemblies are not fiscally autonomous – that is, their finance remains drawn from a central block grant – the fate of many devolved transport strategies and initiatives rests with the decisions of HM Treasury. For example, the promotion of low-cost air travel as a means of enhancing tourist and business access to Scotland and Northern Ireland would probably be seriously undermined if central government decided to raise the taxes on aviation significantly. Equally, the devolved administrations are unable to introduce a number of sustainable transport measures, such as more vigorous taxation of the benefits from running company cars or tax relief on public transport season tickets, because they lack the necessary powers.

Transport Policy Choices from Devolved Perspectives

Chapter 1 demonstrated that there has been a gradual retreat from the sustainable transport policy ideas articulated by Labour in its early period of office. Faced with the renewed political confidence of the 'car-owning democracy', the White Paper, *A New Deal for Transport: Better for Everyone*, failed to fully articulate the case for more sustainable transport in terms of the likely benefits to the economy, environment and society.[14] Yet it was not just Labour at Westminster that realized the introduction of genuinely radical transport policies would be politically difficult. The new devolved governments were also quick to understand that there were very clear local reasons why support for such policies was unlikely to be forthcoming in their own jurisdictions. Indeed, the fact that the new institutions craved popular support to cement their legitimacy probably made them even more sensitive to the demands of articulate, pro-car interests.[15]

An important additional factor is that the very notion of 'sustainability' is perceived differently in different 'regional' or 'national' contexts. To date, British society has been unable to decouple economic prosperity from rising demand for transport.[16] But the political and policy trade-offs that must be addressed between economic prosperity and environmental challenges differ markedly across the UK. Perhaps the most significant issue is that political and public opinion in the devolved countries does not perceive transport related problems to be as serious as is the case in England. This is partly because congestion is (on the whole) less severe, but also because the political influence of rural interests – many of whom promote *more* car use as a means of stimulating local economies in areas often difficult to serve by public transport – is relatively much stronger.[17] Being generally less prosperous, and with large rural areas with very low population densities, the priorities of policy makers in Scotland, Wales and Northern Ireland therefore tend to focus much more on regional economic development, employment creation and social inclusion rather than the problems associated with rising traffic (Table 2.2). Under these circumstances, it is understandable that a locally sensitive definition of sustainability might accept that a degree of traffic growth is necessary to secure the economic and social sustainability of peripheral regions.

Further divergence between Westminster and the devolved administrations is apparent in their approach to international transport connections. This difference in priorities was reflected in the lack of emphasis which was given in *A New Deal for Transport* to air transport and ports, which are

Table 2.2   Population characteristics of the United Kingdom

	England	Scotland	Wales	Northern Ireland	United Kingdom
Population (000s)	49,997	5,115	2,946	1,698	59,746
Population density (people per km^2)	383	65	142	125	246

Source: Office for National Statistics (2002) *Britain: an official handbook*. The Stationery Office, London.

regarded as critical to the future well being of Scotland and Northern Ireland among opinion formers and decision makers in those areas. In particular, the varying propensity for air travel across UK regions – Scotland generates the second highest level of air travel per capita, while Northern Ireland is a close third – reveals the extent to which the role of different transport modes is perceived differently in different socioeconomic contexts. Put simply, in both Northern Ireland and (to a lesser extent) Scotland, peripherality combined with the absence of competitive alternative modes makes business travel to the core of the UK economy in the south east highly dependent on air services. Maintaining scheduled links to the UK's major international hub at Heathrow is also crucial, with Westminster's procedures for the allocation of air slots in London already becoming an issue of contention for the devolved administrations (Chapter 9).

Transport Policies and Strategies in the Devolved Nations

The conflicting pressures of enhancing regional economic prosperity whilst minimizing the environmental impact of air travel epitomize the wider challenge of achieving a sustainable transport policy across the devolved UK. The devolved administrations are intimately aware of their distinctive political contexts, and have developed policy strategies which, although autonomous, *de facto* mirror the retreat from the sustainability agenda seen at the UK level, almost as if Westminster's policy conclusions had been adopted through a process of osmosis. Thus it appears that the increased discretion to meet local demands afforded by devolution provides another set of reasons for government to back away from implementing more sustainable transport strategies.

Scotland

In 1998 (a year before devolution came into effect), the former Scottish Office published its own Transport White Paper, *Travel Choices for*

Scotland.[18] Following 20 years when Scotland did not have a coherent strategy for transport, the government set out to adopt a more sustainable agenda by shifting resources and attention to the provision of noncar modes. The White Paper foresaw better integration between different transport modes, and improved coordination between transport and other policy areas.[19]

Following devolution, the new administration addressed the issue of sustainability in the Executive's first *Programme for Government* published in 1999.[20] The document recognized the need to deliver distinctive solutions that reflected the diversity of issues in different parts of Scotland. It developed a number of policy themes including promoting modal shift, reducing traffic growth and congestion, increasing support for public transport and improving 'lifeline' services to islands and other remote areas, which were consolidated in the Transport (Scotland) Act (2001).[21] Mirroring experience at Westminster (Figure 3.1), the coherent development of transport policy in Scotland has been hampered by consistent change in the structure of government departments and the appointment of succession a transport ministers reminiscent of a 'revolving door' situation (Table 2.3).

The Scottish Executive now provides funding for the majority of transport in Scotland, including the motorway and trunk-road network, the ScotRail passenger rail franchise, ferries, Highlands and Islands Airports,

Table 2.3   Changes in responsibilities for transport, planning and environment portfolios in Scotland

Date	Minister	Portfolio
November 1999	Sarah Boyack	Minister for Transport and the Environment (including planning)
November 2000	Sarah Boyack	Minister for Transport
	Sam Galbraith	Minister for Environment (including planning)
May 2001	Sarah Boyack	Minister for Transport and Planning
	Ross Finnie	Minister for Environment and Rural Development
November 2001	Wendy Alexander	Minister for Enterprise, Transport and Lifelong Learning
	Iain Gray	Minister for Social Justice (including planning)
May 2002	Iain Gray	Minister for Enterprise, Transport and Lifelong Learning
	Margaret Curran	Minister for Social Justice (including planning)
	Ross Finnie	Minister for Environment and Rural Development

Source: Allmendinger, P (2002) *Integrating planning in a devolved Scotland.* Paper presented to the Planning Research Conference, University of Dundee, April.

Table 2.4   Planned transport expenditure in Scotland, 2002

Inland surface public transport £m		Self-propelled transport £m		Roads £m	
Public Transport Fund supports 79 projects to include: (1) Quality bus corridor in Glasgow (2) Park & Ride and bus priority measures in the North East (3) Eriskay Causeway (4) Light Rail System in Edinburgh (5) Rail passenger subsidy for ScotRail	175.0	Walking & cycling and Safer Streets Projects, Scotland-wide scheme	20.0	Motorways and trunk roads 94 schemes throughout Scotland: Completion of M74 New building on A1, M77, Glasgow Southern Orbital Road and A830, A90 grade separated junction improvements	320.0

Other targeted areas

Freight Facilities Grant scheme	March 2002 target achieved of transferring 18 million lorry miles per annum off Scotland's roads to rail and water. £37.6m awarded to 14 companies. Targeting a further 3 million lorry miles per annum – to be transferred by March 2003.
Life line ferry services	£21.4m for CalMac in 2002. Funding has been committed for two new vessels. From October 2002, Northlink Orkney and Shetland Ltd have served the Northern Isles with three new ships.
Life line air services	Highlands and Islands Airports Ltd. To receive £91.4 million between April 2001 and March 2004. New terminal buildings are secured at Inverness, Kirkwall and Stornoway. Runway improvements secured at Wick and Benbecula.
Rail passenger services	Improvements made through the public and integrated transport funds. Awards include: investment in the Fife Circle Line (£8.22m), new trains for Strathclyde Passenger Transport (over £13m), over £24m towards the proposed Larkhall to Milngavie rail link, and Park and Ride facilities and station improvements at Falkirk High (£3.95m)

Sources: Scottish Executive (2002) *Scotland's Transport: Delivering Improvements.* Scottish Executive, Edinburgh; Scottish Parliament (2002) *Report on Stage 1 of the 2003-2004 Budget Process,* Transport and Environment Committee, session 1, http://www.scottish.parliament.uk/official_report/cttee/trans-02/trr02-budgetprocess-01.htm (accessed 15 December, 2002); Scottish Parliament (2002) *Report on 2003-04 Budget Process at Stage 2,* Transport and Environment Committee, session 1, http://www.scottish.parliament.uk/official_report/cttee/trans-02/trr02-stage2budget-01.htm (accessed 15 December, 2002).

Bus Service Operators' Grant (Chapter 7), road safety and Freight Facilities Grants. Direct Executive spending – which excludes the local road network since this is funded by councils – is planned to more than double in real terms to £586 million by the end of 2003/04 (Table 2.4). Of this, spending on the Motorway and Trunk Road Programme has remained about the same at £224 million. At first glance, spending on public transport appears impressive, increasing from £32 million in 1996-97 to £363 million by 2003/04 – almost twelve times as much in real terms. Closer inspection, however, reveals that this increase is almost entirely due to the transfer of accounting responsibility for ScotRail from the SRA to Holyrood (Figure 2.1).

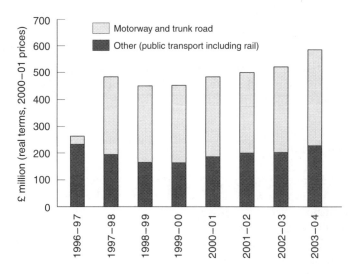

Figure 2.1 Investment in motorways / trunk roads and other transport modes in Scotland, 1996/97 – 2003/04. Source: Scottish Executive (2002) *Scotland's Transport: Delivering Improvements.* Scottish Executive, Edinburgh.

Wales

The Welsh Assembly Government's transport policy statement promised progress on a number of fronts, including a trunk roads review, improved public transport, new powers for local authorities to introduce road user charging and workplace parking levies, innovative cycling and walking initiatives and the transfer of more freight from road to rail. All Welsh local authorities published Local Transport Plans (LTPs) during the summer of 2000, with Bus Subsidy Grant and Welsh Rural Transport Grant combined under the control of local authorities to promote integration (Chapter 3). In January 2001, the National Assembly announced a £300 million package to support local transport delivery over the following five years. The measures provide for the implementation of a number of road improvement schemes, rail infrastructure improvements and £23.2 million dedicated to integrated transport schemes.

The *Transport Framework for Wales*, published in November 2001, sought to provide the basis for a co-ordinated approach to transport, which would integrate the economic, environmental and social objectives of local authorities, transport operators and the National Assembly (Table 2.5).[22] Whatever the outcomes turn out to be, much of the wording of the promised action plan is vague and unspecific, being open to varying interpretations by those responsible for policy delivery.

Northern Ireland

It is widely perceived that almost three decades of direct rule over Northern Ireland from 1972 led to substantial underfunding of the Province's transport system.[23] However, the recently created Department for Regional Development (DRD), which looks after transport, has been a radical and active Ministry, and is regarded by many commentators as one of the success stories of devolution. A Railways Task Force exerted a huge influence on the debate over the future of Northern Ireland Railways (NIR). The Task Force initiative had been prompted by an internal report that raised fears that most of the service might have to be closed down.[24]

When the Task Force's report was published in September 2000, the Northern Ireland Executive was persuaded not to recommend closure of the Province's rail network, an option that had apparently been favoured by the previous direct rule minister. Its plan envisaged the consolidation of the existing network with priority for the most heavily used lines. In due course, the Northern Ireland Transport Holding Company, the parent company of

Table 2.5   Transport expenditure breakdown in the Welsh 'Five-Year Plan'

Improvements to public transport	Encourage rail operators to carry out improvements and encourage improved interchange, timetabling, information and ticketing. Devote an increased proportion of the transport grant funding for local authorities to public transport projects.
Action for rural areas	£2.25m per year for new and improved rural bus services, promoting accessibility.
Road safety	£250m a year for community transport; up to £500,000 a year will be made available through the Transport Grant Arrangements for the New Safe Routes to Schools Initiatives.
Trunk roads	A strategic review of the Welsh Trunk Roads Programme – a new core trunk road network has been defined.
Green transport	Introduction of Green Transport Plans by all major employers and revised criteria for dealing with noise pollution in specific areas.
Integrated transport	£23.2 million has been dedicated to integrated transport schemes.

Source: Welsh Assembly Government (2001) *Transport Framework for Wales*. WAG, Cardiff.

the Province's state-owned bus and rail operators, placed an £80 million order for new trains to operate over the domestic network.

The DRD also launched an ambitious Regional Transport Strategy (RTS) in July 2002. The RTS reflected the transportation principles in the Northern Ireland Transport Policy Statement, *Moving Forward*, which set out a strategy for implementing the objectives of the UK White Paper in a way which would reflect the particular circumstances of Northern Ireland.[25] It acknowledged that Northern Ireland suffered from underinvestment in transport compared with England, Scotland and Wales. Although it claimed to signal an important step towards a more balanced and integrated system, in which public transport and nonmotorized transport would be attractive options for many trips, this was not fully reflected in the balance of expenditure: roads would receive 63 per cent of the £3500 million budget over 10 years, and public transport 35 per cent. Table 2.6 summarizes the key elements of current and projected priorities and spending on transport in Northern Ireland.

The suspension of the Northern Ireland Assembly in the autumn of 2002 has left these plans in limbo. It will be for direct rule ministers appointed by London to decide whether or not to take them forward. Past experience has shown that such appointees tend to be cautious of taking controversial

Table 2.6 Strategic transport programmes and spending in Northern Ireland

Public transport £m		Personal modes £m		Roads £m	
Bus	628.5	Walking & cycling	86.8	Strategic roads	681.5
Enhanced facilities;		Traffic calming;			
Fuel duty rebate;		Making it easier to walk;			
Concessionary fares;		Making it easier to cycle;			
Bus replacement		Improving cycle links at 120 sites; Improved accessibility			
Rapid transit	100.7			Roads maintenance (structural): Accident remedial schemes; Traffic management; Routine maintenance; Bridge strengthening; Network management costs	1,499.6
Rail	502.9			Network development schemes; Street lighting; Promoting sustainable modes	
	1,232.1		86.8		2,181.1

Source: Department for Regional Development (2002) *Northern Ireland Regional Transportation Strategy 2002–2012*. DRD, Belfast.

decisions, preferring to leave them to the Assembly. This suggests Northern Ireland could face paralysis in decision making in many areas including transport.

Public Expenditure Differentials: A Barometer of Policy Delivery?

Over and above the current policy agenda and existing conditions, different parts of the UK have spent different amounts per head of population on transport in the past. These inherited trends of spending – which may not even have a fully rational basis and might be hard to alter – nevertheless give a clue as to the different priorities in each jurisdiction. Expenditure on transport can therefore provide a valuable barometer of the extent to which the rhetoric of national policy is converted into action – in other words, how each devolved administration perceives the importance of transport. Yet in comparing the scale of expenditure in the different administrative units of the UK, it is not only the impacts of scale and the Barnett Formula that have to be taken into account. The composition of spending between capital expenditure on new transport infrastructure and revenue commitments such items as road maintenance and public transport subsidies also needs to be recognized.

Overall, spending on transport (both capital and revenue) in the UK is typically significantly less than in most other major EU countries (Chapter 1). HM Treasury provides estimates of identifiable total managed expenditure for England, Scotland, Wales and Northern Ireland and the UK as a whole (Table 2.7). The figures demonstrate that per capita spending in Scotland was higher than that in England, but significantly higher than in both Wales and more particularly Northern Ireland. A review of earlier years highlights the fact that Scotland has consistently achieved higher overall annual total managed expenditure per head than any other part of the UK.

Figures published separately by the Scottish Executive, the Welsh Assembly Government and the DETR on central and local government expenditure on transport also confirm that overall annual expenditure in 2000/01 was higher in Scotland than either England or Wales. But when the overall figures are broken down between capital and current expenditure, the UK statistics reveal that less per head was invested in capital items in Scotland than in either England or Wales during 2000/01, although the Scottish figure was higher than that for Northern Ireland. Moreover, it should also be recognized that transport expenditure across England is skewed by the complexity and size of transport infrastructure in London and the South East, and a more appropriate comparison would be with transport expenditure between comparable metropolitan regions of England.

Table 2.7   Identifiable total managed expenditure per head on roads and transport (£s)

Year	England	Scotland	Wales	Northern Ireland	United Kingdom
1999/2000	133	188	154	123	138
1998/99	144	183	150	120	147
1997/98	154	196	156	120	157
1996/97	167	219	192	124	172
1995/96	193	240	221	125	196

Source: HM Treasury (2002) Personal communication.

A distinct difference, however, applies in relation to revenue spending due to the generous (by UK standards) subsidies given to supporting lifeline air and ferry services and related infrastructure in Scotland. Such provision is almost unique in the UK and although the need is explained by geography, similar arguments could be mustered in the case of Northern Ireland, where almost no such provision is made. The implication of this distribution of public spending on transport is that it obviously curtails opportunities for enhancing other financially or economically more beneficial infrastructure and services, in the central belt for example, given the limitations which clearly apply to the Scottish block grant as a whole. Scotland is therefore generously resourced compared to either Wales or Northern Ireland, but the difference between Scotland and England lies in how the money available is allocated between capital works and recurrent spending.

Promoting a Sustainable Transport Policy: Is Devolution a Help or Hindrance?

It is all too easy to see how local circumstances – such as peripherality and low population density – might provide an additional set of structural difficulties for any devolved administration seeking to advocate a sustainable transport strategy. With the very definition of sustainability perceived differently across the UK, it is also clearly apparent that coordination of transport policy between Westminster and the devolved administrations – if deemed desirable in areas of mutual interest – is likely to be problematic. Put simply, it could be said that a 'one size fits all strategy' for what sustainable transport actually means for the UK will not work. An analogy might be drawn here with the debate on the UK's possible entry to the European Single Currency, with northern exporters and manufacturers being broadly supportive of entry to the Euro while opinion in the London financial centre is generally more sceptical.

Given the way responsibility for transport is divided between Westminster, the devolved assemblies, local authorities, various quasi-public agencies and private sector operators, there is little prospect of an integrated transport policy being led from the centre in any case. What therefore is most surprising about the experience of transport policy development under devolution to date is that a common strategy of sorts exists at all. In short, the various administrations currently share a general vision to develop more sustainable transport, but they are all also similarly politically nervous about making the radical change required in order to deliver these aspirations, particularly given local circumstances. Each of their transport strategies is testament to this.

A UK transport policy beyond Watford?

Viewed from the periphery of the UK, it is evident that the priorities of government officials and many in industry in Northern Ireland, Scotland and Wales differ from the typical resident or policy maker in London and the South East. Continuation of differences in societal values within constituent parts of the UK will ultimately determine whether a sustainable transport policy founded on the principles advanced by Labour will be maintained. This is because policies that might be reasonably regarded as symptomatic of 'retreat' from the aspirations of sustainable transport in the majority of the UK might be entirely rational, even sustainable, given the unique set of economic, environmental and social policy challenges in each particular devolved context. These perceived differences in the significance of competing policy approaches depend not only on local conditions but also on the likely effectiveness of available policy measures. The latter is likely to be of particular importance in prioritizing elements of potentially effective strategies.

One example which illustrates this point well is the desire to achieve a significant modal shift from the private car to public transport. The apparent inexorable rise in car travel in part reflects the failure of existing market mechanisms to account for the true environmental, economic and social costs of car use. This then has the effect of stimulating demand for this mode of travel over and above the level expected if its price accurately reflected its true costs (Chapter 1). The effect of reduced transport costs – in effect hidden subsidies for car use – is eventually to promote the dispersal of activities and places of residence. In much of urban Britain, this decentralization and dispersal is one of the most profound problems resulting from overreliance on the car. But in many remote rural areas, particularly in the devolved countries, *sustaining* economic and population dispersal is crucial to maintaining the viability of remote communities. Indeed, it could even be argued that this is entirely consistent with the UK government's own decision to widen the definition of 'sustainability' in *A New Deal for Transport* sufficiently to include almost all of its other policy initiatives ongoing at the time.

Moreover, it is not the absolute level of service which determines modal share but the differences between competing alternatives. Public transport can only compete effectively where it offers comfort and either a substantial in-vehicle journey time or access/egress time advantage over the car. Such a situation will occur where there is serious congestion and/or nearby parking is not readily available, or is expensive, and public transport is able to avoid the effect of congestion, by being segregated from the private car. It is in

those areas with comprehensive and effective public transport allied to significant traffic congestion or parking restraints that the difference in general-
ized costs is closest, and thus behaviour is most sensitive to change with less
radical measures or relatively lower levels of transport expenditure on a per
capita basis. This is particularly true for central London, and was reflected in
the successful launch of the congestion charging scheme in February 2003
(Chapters 4 and 10). In contrast, in most provincial cities, public transport
faces a bigger challenge in winning market share as well as retaining it. This is
even more apparent in smaller centres and particularly in rural areas. In these
cases it is quite hard to envisage circumstances where, given a choice between
the bus and car, the former can compete successfully.

Against much conventional wisdom, therefore, it might be that although
the potential for achieving a transfer from car to public transport appears at
first sight to be greatest in areas of highest car dependency, in practical
terms, the extent to which non-car modes can represent realistic alternatives is open to some serious questions. Chief amongst these is whether the
political will exists to meet the likely adverse public reaction to the pricing
measures and/or the (relative) scale of the funding required to bring about
this degree of change in those areas. Thus it may not be unexpected to find
devolved policies developing at odds with one of the key principles of
sustainable transport, particularly in rural Scotland and Northern Ireland,
where car dependency is highest in the UK.

It is tempting therefore to conclude that devolution can only coexist with
sustainable transport policies in a weak and fragile way. In some ways this is
not surprising, since the point about devolution is that it allows local
politicians to devise local solutions for local problems *as perceived in the
devolved areas*, rather than having to implement a UK-wide policy handed
down from the centre. But this would be to discount the latent potential of
devolution, since it is equally possible that, given time, devolution could
prove to be entirely positive for the promotion of sustainable transport
ideals, if the political barriers to their implementation can be overcome.
Indeed, if one devolved administration was to pioneer the delivery of such
policies, this might come to represent the leadership required to convince
other governments to follow suit.

Even if devolution turns out to be an impetus for more sustainable
transport policies in some areas, there will still need to be effective forums
for discussion between UK administrations to formulate agreed positions
on particular issues over which the influence of reserved transport powers is
especially important. Arguments could be put forward for joint sessions of
elected representatives and members of the respective devolved executives
to promote the coordination in specific policy areas of mutual interest.
Whether contentious issues can be resolved – such as the future of slots
at key hub airports including London Heathrow that are currently allocated

to links to Scotland and Northern Ireland – will be a key test of central government's commitment not to use its reserved powers to promote the interests of London and the South East at the expense of the devolved areas. There may also be merit in undertaking comparative studies of how formerly/currently eligible peripheral areas have been able to exploit EU funding opportunities and how effective such finance has been in contributing not only to transport but wider social and economic goals.

It is also possible that coordination between Westminster and the devolved areas (and in many cases also the Republic of Ireland) could usefully extend to harmonization of fiscal and other financial measures, market regulation issues, and joint strategies for transport infrastructure and service provision. Indeed, the British-Irish Council (BIC), in which the Irish Government, UK Government and devolved administrations discuss issues of common interest, regards transport as being one of the policy areas for which cooperation across the British Isles could be most beneficial. In the longer term, therefore, if the political will exists, it could be that devolution turns out to be complementary to the achievement of sustainable transport aims. But, as for UK central government, the devolved administrations will have to avoid the temptation to formulate their transport strategies according to political expediency.

NOTES

1   The UK government in London retains responsibility for 'reserved' powers (including macroeconomic management, foreign affairs and some domestic matters such as transportation safety) and English domestic administration.

2   Department of the Environment, Transport and the Regions (1998) *A new deal for transport: better for everyone*. Cmnd 3950, The Stationery Office, London. Prior to devolution, separate government departments, The Scottish Office, The Welsh Office and the Northern Irish Office, existed for the administration of the three countries. Ministerial representation in these departments was drawn, like any other, from MPs of the governing party at Westminster.

3   *Western Mail* (2000) 6 May. In Scotland, devolution was supported by 74.3 per cent of those voting in the referendum, while in Wales the proportion voting 'yes' was only 50.3 per cent.

4   The Northern Ireland Assembly was suspended for over 100 days during the first half of 2000. It was suspended again indefinitely on 14 October 2002 amidst allegations of bad faith by Unionists against Republicans and vice versa.

5   Hazell, R (ed) (2000) *The state and the nations*. Imprint Academic, Exeter.

6   For more information on the powers and responsibilities of the new institutions of London government, see www.london.gov.uk/approot/about.jsp (accessed 15 December 2002).

7   Transport (Scotland) Act (2001) The Stationery Office, Edinburgh.

8   See the evidence to the Scottish Parliament's (2002) *Report on inquiry into the rail industry in Scotland*, Fifteenth Report of the Transport and Environment Committee, Session 1, for a range of views over how the limited extent of the rail policy powers granted to the Scottish Executive impacts on the management of the railways in Scotland. www.scottish.parliament.uk/official_report/cttee/trans-02/trr02-15-01.htm (accessed 15 December, 2002).

9   Glaister, S (2001) *UK transport policy 1997–2001*. Address to the British Association for the Advancement of Science, Glasgow, September 4, 30.

10 Tomaney, J (2000) *The governance of London*. In Hazell, R (ed) *The state and the nations*.

11 For the full list of powers reserved to Westminster under Scottish devolution, see The Scotland Act (1998) *Public General Acts – Elizabeth II*. Chapter 46, The Stationery Office, London.

12 For a comprehensive review of current perspectives on Scotland's fiscal position within the UK, see *Scottish Affairs* (2002) Special Edition on Fiscal Autonomy, autumn.

13 HM Government (1998) *The Barnett formula: the government's response*. Appendix 2, HC 619, 10 March, The Stationery Office, London.

14 DETR (1998) *A new deal for transport*.

15 Docherty, I (2000) Rail transport policy-making in UK passenger transport authority areas. *Journal of Transport Geography*, 8, 157–70.

16 For more on the Chair of the Commission for Integrated Transport's view on decoupling traffic growth from economic growth, see Begg, D (2001) Hit the brakes, *The Guardian*, 6 March.

17 Gray, D; Farrington, J; Shaw, J; Martin, S and Roberts, D (2001) Car dependence in rural Scotland: transport policy, devolution and the impact of the fuel duty escalator. *Journal of Rural Studies*, 17, 113–25.

18 Scottish Office (1998) *Travel choices for Scotland*. The Stationery Office, Edinburgh.

19 Docherty, I and Hall, D (1999) Which travel choices for Scotland? A response to the government's White Paper on integrated transport in Scotland. *Scottish Geographical Journal*, 115, 192–209.

20 Scottish Executive (1999) *Programme for government*. The Stationery Office, Edinburgh.

21 Transport (Scotland) Act (2001), The Stationery Office, Edinburgh.

22 Welsh Assembly Government (2001) *The transport framework for Wales*. WAG, Cardiff.

23 'Direct rule' refers to the situation when domestic government for Northern Ireland is exercized by ministers of the UK government.

24 Department for Regional Development (DRD) and Northern Ireland Transport Holding Company (2000) *Interim report of the railways task force on the future of the railway network in Northern Ireland*. DRD, Belfast, September.

25 Department of the Environment for Northern Ireland (1998) *Moving Forward*. DoENI, Belfast.

3

Local Transport Planning under Labour

Geoff Vigar and Dominic Stead

The opening chapter highlights the rather limited extent to which Labour has changed the direction of UK transport policy since 1997. Labour started with radical rhetoric, but its vision was increasingly watered down as it responded to popular protest about motoring costs and the perceived and actual absence of alternatives to the car for many people. Despite the assertion of the 1998 White Paper, *A New Deal for Transport: Better for Everyone*, that the public wanted change and was ready for the sort of radical policies the government was proposing, even its reduced measures can latterly be seen in political terms as 'too much too soon'.[1] This chapter seeks to explore the changes to transport planning within local authorities, who are crucial players in the development and implementation of transport policy. Whatever central government's rhetoric, it relies heavily on local authorities to deliver much of its agenda for transport, both in bringing forward ideas and in carrying forward their own and central government's policy principles into the detail of local policies and proposals.

This chapter is structured into four sections. The first focuses on the role of local authorities in transport policy and provision, and the changes since 1997 in these areas. The second section expands on this introduction and proposes four themes that characterize Labour's approach to governance and transport policy at the local level (these themes also have a wider application beyond the specifics of this chapter): the pursuit of *integration*, varying forms and degrees of *decentralization*, a somewhat greater belief in the power of *public spending* than its predecessor Conservative administrations, and commitments, rhetorically at least, to *participation*. The section looks at the key legislative and policy material driving these themes and how local authorities have responded to this agenda, before drawing attention to the key areas of conflict between national and local policy priorities. The third section examines the progress and continuing difficulties experienced by local authorities in transport planning, drawing partly on primary research conducted during Labour's first term of office. In the final section, we conclude with some remarks about future local transport policy given Labour's difficulties in pursuing its transport agenda.

Changes to Local Authority Responsibilities

In most of England little has changed in terms of local government's formal responsibilities for transport planning. County, metropolitan and unitary councils remain the designated 'highway authorities' responsible for the maintenance of the road network beyond a core of trunk routes (Chapter 4). Such authorities are also responsible for promoting an integrated transport agenda, including subsidising public transport fares and services where considered appropriate. In England's six former metropolitan counties, borough councils remain as members of Passenger Transport Authorities (PTAs) charged with coordinating public transport and maintaining and operating some other local services and infrastructure such as ferries and tunnels. Significant change has, however, occurred in terms of responsibilities in London. Here the Mayor, overseen by the Greater London Assembly (GLA), has a new responsibility for producing a transport strategy for London and has gained control over most public transport fares, excluding main-line rail services (Chapter 6).

In terms of the process and substance of policy making, there is some considerable change to report since 1997. Local government has been granted control of more of the road network as central government has proceeded with its policy of 'detrunking' less heavily used routes. In addition, the role of regional planning institutions has been strengthened from a purely advisory/consultative role to being responsible for producing *Regional Planning Guidance* (RPG) and *Regional Transport Strategies* (RTSs) in consultation with Regional Development Agencies and other bodies.[2] This means that local authorities remain important as key shapers of local transport policy in their own right, and also as members of Regional Planning Boards. The Transport Act (2000) also introduced a range of new policy tools. Chief among these were powers for local authorities to introduce congestion and workplace parking charges (Chapter 4) and the ability to enter legal agreements over Quality Partnerships and Contracts with bus operators (Chapter 7). The Road Traffic Reduction (National Targets) Act (1998) has also put an onus on local authorities to address traffic growth and to have a strategy for considering it (Chapter 4). Similarly, they are now required to have a cycling strategy (Chapter 8).

Labour has also changed the ways local transport is funded. The former annual Transport Policies and Programmes statements have been replaced

Table 3.1   Main differences between the TPP and LTP systems

Transport Policies and Programmes (TPP)	Local Transport Plans (LTP)
Produced annually	5 year plans; greater certainty of future funding for authorities
Primarily a bidding document for central government funds	Partly a bidding document, but also a strategic planning document for a local audience
Programme of capital investment	Consideration of both capital and revenue spending
Resources tightly ring-fenced to particular areas of expenditure	Greater local discretion over allocation of resources
Limited input from operators and other local partners	Inclusive approach, involving operators, greater public and local business participation
Broad objectives (except under package approach)	Greater emphasis on targets, performance indicators and monitoring in areas not previously covered by packages
Historic emphasis on road schemes, although shift in recent years to measures encouraging use of noncar modes.	Emphasise integrated transport solutions to encourage public transport, cycling and walking.

Source: Department of the Environment, Transport and the Regions (2000) *Guidance on Full Local Transport Plans*. DETR, London.

by Local Transport Plans (LTPs) in England and Wales, and Local Transport Strategies (LTSs) in Scotland (Table 3.1). Under the Transport Policies and Programmes (TPP) system, local authorities requested government funding for individual schemes. This was inefficient not only in terms of time and resources, but it also meant that decisions were taken in isolation, and often made on purely financial grounds, rather than considering their contribution to a strategy. The LTP/S system, in contrast, is built around five-year integrated transport strategies.[3]

Local Transport Planning: Key Themes, Changes and Responses

Integration

There is a long history of (physical) integration in the transport field that seeks to improve interchange between modes (intermodality) and the interoperability of transport systems.[4] Such issues have not gone unnoticed by Labour, and indeed *A New Deal for Transport* mentioned the need to develop the capacity for such integration a great deal. Labour has, however, focused greater attention both within the transport field and beyond on the theme of *policy* integration, or 'joined-up government', to coin a New Labour expression. This would appear to be a pet project on the part of Prime Minister, Tony Blair, and as such has driven much of what government has done during its first and second terms. Policy integration attempts to put service users at the heart of governance attention. It is argued that policy areas such as transport can be trapped in a 'silo mentality' that focuses on administrative convenience rather than users of a service.[5] It also reflects the fact that activities in policy sectors often conflict with objectives in other sectors and that meeting transport objectives, for example, critically depends on policies and decisions taken in other policy areas.[6] Communicating transport policy aims to other policy communities then becomes of critical importance. The metaphor of 'joined-up thinking' has been deployed as a call for public sector stakeholders in particular to continually look to other policy communities when making decisions and policy.

We can thus identify two types of policy integration. One is vertical policy integration, which refers to integration between different levels of government (national, regional and local). A second is horizontal policy integration, which refers for example to integration between different policy communities at a given spatial scale. This can imply relations between a local authority and other stakeholders or between departments in the same organization. A subset of horizontal integration refers to intradepartmental integration, where coordination occurs between individual departments that are split, perhaps in terms of organizational boundaries (for example, separate sections or units) and / or functional boundaries (for example, according to transport mode).

Vertical policy integration in terms of the relations between central and local government is dominated by central government's control of local authority transport policy, principally through approval of financing. This potentially maintains a large degree of integration over policy objectives as central government can exercise a gatekeeper role over such funds. We discuss changes to this relationship in more depth in the 'decentralization' section below.

In England, the region has been increasingly used as a way of integrating both policy agendas (horizontal integration) and getting beyond problems associated with central–local government relations (vertical integration). A further key integrating force for planning and transportation communities has been the creation of the RTS. This elevates regional transport policy to greater standing than under the previous RPG preparation process, although they are to be prepared together, encouraging the integration of the two policy agendas. Certainly there is evidence that this is giving transport issues a greater voice over planning concerns such as the designation of large employment and housing sites.[7] However, concerns remain about the institutional mechanisms that exist at regional level and whether meaningful strategy can be developed without greater accountability, such as elected regional government of some kind.[8] Indeed, Regional Planning Guidance is not legally binding on development plans and LTPs.[9] Government proposals to unite the two documents into a Regional Spatial Strategy (RSS), with a greater degree of freedom for regions to set priorities than under RPG, promotes integration at the regional level, although this is offset somewhat by the abandonment of statutory structure plans which have proved to be a key integrating device and also a useful opportunity for the public and other stakeholders to participate in transport policy making in the past.[10] In addition, the preparation of LTPs in advance of RTSs, and the timing of the multi-modal studies (MMSs), has to date prevented the Strategies from providing a clear framework in which to plan and implement best policies (Chapter 4).

The second form of integration, horizontal integration, has often focused on uniting the two policy communities of land-use planning and transport planning, although increasingly links are being made between transport policy communities and those concerned with others such as health. We focus here on links with land-use planning as it is planning policy that consistently has a major bearing on whether transport policy objectives may be met, and to some degree frames the spatial ambitions of other policy sectors and thus their long-term transport needs and requirements.

In recent years, there have been signs of attention to this agenda. For example, under John Major's Conservative government of 1992–97, the Department of the Environment (responsible for planning matters) and the Department of Transport commissioned research on a joint basis and also produced a collaborative statement of planning policy guidance concerning transport.[11] In 1997, one of Labour's first acts on taking office was to merge the Department of the Environment (responsible for planning) and the Department of Transport to create the Department of Environment, Transport and the Regions (DETR) in an attempt to 'join-up' these policy areas. Obviously, there is more to policy integration and breaking down the 'silo mentality' than simply restructuring and renaming departments. As the Royal Commission on Environmental Pollution (RCEP) noted, 'restructuring government departments does not in itself guarantee that coherent policies will emerge'.[12] What happened after the formation of the DETR (and some local government departments that restructured in a similar way) was that planning and transport responsibilities initially remained in separate sections and even in separate buildings.[13] Although this situation changed over time as a result of office moves and staff mobility between sections, professional boundaries remain important in demarcating responsibility for policy making in both central and local government.[14] For day-to-day activity, the two policy communities are largely answerable to different constituencies and attention within transport has been diverted to focus on higher profile issues such as delivering improvements to rail services, perhaps to the detriment of integration efforts. Within the space of a few years, the DETR 'super-department' proved too unwieldy and unmanageable and has undergone reorganization twice since 1997 (see Figure 3.1). In Scotland, departmental responsibilities for transport, planning and environmental policy have been subject to even more change than in England (Table 2.3).[15]

Long-term decision making is crucial to the integration of transport and planning policy and the creation of both RTSs and the five-year LTP has encouraged greater attention to transport strategy-making. Links to planning are typically better developed than under the previous LTP system.[16] The requirement to produce Transport Assessments of major developments also provides an opportunity for transport and land-use agendas to come together. How far such assessments are able to challenge the

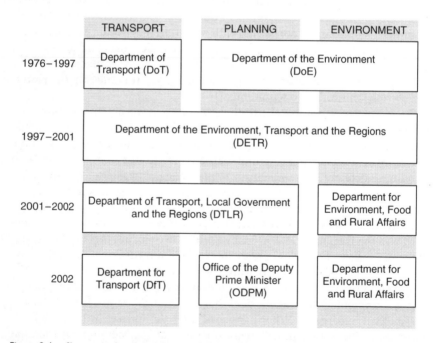

Figure 3.1   Changes in departmental responsibilities for transport, planning and environment policy in England.

'presumption in favour' of development within the planning system on transport grounds alone is, however, a moot point. The ability of Transport Assessments to promote policy integration between neighbouring authorities is also limited (see below).

Where unitary authorities have been created (all of Scotland and Wales, and many major towns and cities in England), it might seem logical that decisions should now be more integrated because coordination of decision making should be easier with fewer tiers. In reality, however, there have been significant problems of spatial integration and coordination across authorities.[17] Strategic transport planning is problematic where there is now no coordination at the county level. Whilst acknowledging that the search for a best fit of travel patterns into a designated area is something of a lost cause, new unitary authorities are often small and lack the scale to develop meaningful strategic transport planning. Such problems were foreseen well before these local government changes. The 1997 report of the RCEP asserted that 'the area covered by a unitary authority bears too little relation to travel patterns to make a sensible unit for drawing up and implementing integrated transport plans'.[18] Furthermore, the division of responsibility into unitary authorities without a strategic overview can result in development that has (intended or unintended) spillover or knock-on effects for adjacent authorities as a result of too narrow a focus in planning. Transport assessments, now required under Planning Policy and Guidance (PPG) note 13, will not always identify such effects. Thus, in some ways at least, the creation of unitary authorities may have led to less 'joined-up' thinking and less integration of decision making, particularly at the strategic level.

Horizontal integration can also take the form of joint working between local authorities in the production of policy documentation. Whilst such proposals remain nonstatutory, many authorities, particularly unitary authorities, have embraced and often promoted this agenda and sought to develop transport plans jointly with other areas to better reflect the travel patterns of the area (Table 3.2). Such integration at subregional and regional levels has proved increasingly important due to demands for consistency with regard to demand management frameworks.

Despite various mechanisms introduced by Labour, such as the LTP and the RTS, there are concerns that many of the changes have not gone far enough in promoting integration either between modes or with other policy areas. The introduction of MMSs provides a potentially useful mechanism to look at movement patterns and future possibilities without focusing on one mode as has happened in the past. However, the connection to land-use planning appears to have been limited in many instances and many have criticized such studies for quickly resorting to road improvement options with little attention given to pursuing alternatives within a stronger

Table 3.2   Examples of possible arrangements for joint working

Situation	Example
In areas where two or more unitary authorities preparing individual unitary development plans (UDPs) intend to produce a joint local transport plan.	Each unitary authority should include the transport and land use strategy in Part I of their UDP as well as relevant land use planning policies in Part II of their UDP.
In areas where two or more unitary authorities preparing a joint structure plan intend to produce a joint local transport plan.	Transport and land use strategy is included in joint structure plan, and relevant policies in either structure plan or local plan.
In areas where two or more unitary authorities prepare a joint structure plan but decide to produce individual local transport plans.	Individual local transport plans will still need to show how they integrate with those of neighbouring authorities. Joint structure plan should include the transport and land use strategy.
In an area with a unitary authority preparing a UDP and a residual county preparing a structure plan, where the two authorities decide to produce a joint local transport plan.	The transport and land use strategy should appear in both Part I of the unitary authority's UDP and in the structure plan.
In an area with a joint structure plan prepared by one or more unitary authorities and a residual county, where the intention is to produce a joint local transport plan.	The transport and land use strategy should appear in the joint structure plan.

Source: Department of the Environment, Transport and the Regions (1999) *Planning Policy Guidance 12: Development Plans*. The Stationery Office London.

demand management-led policy framework (Chapter 4). In many cases, outputs certainly have tended to favour 'schemes' rather than proposals which might call other policy communities to account or promote whole system solutions instead of new pieces of infrastructure. This often reflects a political realism of what is possible in the short and medium term, but has the problem of maintaining both a culture of car dependency and hyper-mobility and also reflects a degree of path-dependence for local and national transport policy. Furthermore, the preparation processes for many MMSs have proved somewhat opaque, with examples of consultation but a distinctly technocratic feel to many of them.

This failure on the part of many MMSs has been particularly disappointing given the acknowledgement in *A New Deal for Transport* that 'place matters', which implies a particular emphasis on improving quality of life through tackling the impacts of traffic on neighbourhoods. Indeed, this analysis has disappeared almost altogether with the focus on schemes and investment in the *10-Year Plan*.[19] Local authorities' attention has thus been pulled in different directions in a short space of time as they have been forced to shift from the demand management perspective emphasized by the White Paper to the infrastructure development perspective favoured in the *Plan*.

Decentralization

The second theme has been that of decentralization. Local authorities have been given greater responsibility in relation to policy making and implementation. The detrunking of many 'A' roads has given them greater authority over road networks. However, potentially the most significant decentralization initiative has been congestion and workplace parking charging. The Transport Act (2000) enables local authorities to charge for road use in their areas or to place a charge on workplace parking places. Local authorities are allowed to keep the revenue from such schemes for up to 10 years. Whilst this is an interesting piece of enabling legislation that leaves local areas to decide upon whether charging is an appropriate part of a local policy package, it also has helped central government to 'pass the buck' on road pricing (Chapters 1 and 4). Many of the same political difficulties arise locally as nationally when it comes to implementation (see below). To this end, while around 35 local authorities have expressed an interest in implementing a charging scheme, significantly fewer schemes are likely to come forward in the period to 2010 than the 20 originally anticipated by Labour.[20] This undermines both the investment available to local authorities for transport initiatives and also targets for road traffic reduction and congestion.

Greater degrees of freedom as to how to spend their revenues, partly achieved through changes to funding mechanisms, particularly through the introduction of the LTP highlighted above, also illustrate the extent of devolution under Labour. The creation of the 'Single Capital Pot' has also given local authorities greater autonomy and allowed a greater degree of switching of investment between policy areas.[21] However, on-going investments that deliver little political capital in the short-term, such as road maintenance, have sometimes suffered under such arrangements.[22]

Public spending and monitoring

In its first term of office, Labour agreed to maintain the previous Conservative government's spending commitments and levels for its first two years in power. Subsequent to this period it has begun to direct public spending towards priorities more commonly associated with past Labour governments. The *10-Year Plan* assumed £59 billion for 'local transport' in England made up of £30.6 billion public revenue expenditure, £19.3 billion public capital investment and £9 billion private investment. The 2000–01 local transport settlement allocated the first five years worth of the £19.3 billion public investment. The period 2000–05 would see £8.4 billion of government investment to be spent on target projects identified at local level. Of the total, £4.4 billion was made available for public transport, integrated transport and smaller schemes and up to £4 billion for spending on new and existing roads, of which £3 billion was for maintenance and £1 billion for major road schemes. In the 2000–01 settlement funding was approved for 39 road schemes valued at more than £5 million each. These were aimed at providing 'safety and regeneration' benefits. Of 23 new major projects approved in the 2001–02 settlement, 15 were for road schemes. These were justified as being either for 'regeneration purposes' or to relieve traffic levels in towns and villages. This large investment, in bypasses and 'relief roads' in particular, highlighted a shift from the demand management emphasis of *A New Deal for Transport* to a subsequent focus on investing in infrastructure to increase its supply (Chapter 4).

As in various other policy areas, Labour has suffered the embarrassment of underspend in transport (Chapter 1). Following long periods of tough spending constraints, much of the administrative capacity available locally to allocate additional funding has been lost. The situation was exacerbated because Labour also introduced elaborate monitoring systems which further stretched limited professional resources. This emphasis on auditing was also a feature of the Conservative administrations of the 1990s but the extent to which performance indicators were imposed on local authorities through the Best Value inspection regime was highly significant.[23] This regime was accompanied by requirements to monitor a range of other information in LTPs. Greater freedom in funding specific projects was thus accompanied by an increase in assessments at a more strategic level. Future funding was likely to be significantly linked to past performance.

Participation and partnership

A fourth theme of the Labour administration is participation and partnership. Local authorities in particular have been exhorted to engage with a wide range of groups in policy making both in transport initiatives specifically and more generally in areas where transport is a significant issue, such as in the preparation of Community Strategies. The LTP is one case in point where consultation processes have to be built in to the preparation process. *A New Deal for Transport* states that 'we will expect local authorities when preparing their LTPs to consult widely and involve their communities and transport operators in setting priorities for improving transport'.[24] The extent to which such initiatives have been taken on board by local transport policy communities is variable, however. Research suggests that while the letter of such requirements is being followed, some authorities are consulting less than others and many practitioners are failing to see the benefits of developing policies in these ways.[25] There have been problems of bias in such participative efforts also that may serve to skew policy in particular directions. For example, in the preparation of RTSs in England, certain interests, such as the business community, have embraced such strategies and participated fully. Other interests, such as those representing social and health agendas, have been conspicuous by their absence despite efforts to include them.[26]

In terms of more day-to-day partnership arrangements, one interesting example occurs in relation to local bus provision where local authorities can engage with public transport operators on a contractual basis through Quality Partnerships formalized through the Transport Act (2000). Some success in relation to the development of partnerships between transport operators and local authorities in the 1990s in growing ridership on certain corridors led to calls to formalize such arrangements and give local authorities (and PTAs) greater power to enter and enforce such agreements (Chapter 7). — *state not looking at this in diss*

Opportunities, Difficulties and Potential Futures

In previous work four groups of difficulties have been discussed in relation to implementing transport policy: financial, cultural, organizational and political.[27] This four-fold categorization is used in this section to explore the issues facing Labour and local authorities as they strive to implement transport policy. There are, of course, considerable overlaps between these somewhat arbitrary groupings of factors, but it is hoped their presentation here will help efforts to come to terms with what is often perceived as implementation failure in relation to local transport planning.

Finance

Many of the financial difficulties associated with delivering improved local transport systems and improving the qualities of places have been overcome in recent years as both the levels of finance and delivery mechanisms have changed to allow greater flexibility over what, when and how local authorities can spend transport monies. That said, there are concerns about whether local authorities, like central government, can deliver the aspirations contained in the *10-Year Plan*. This difficulty relates to failures to implement charging schemes locally and whether the level of private sector investment anticipated in the *Plan* actually materializes. It also depends on whether local authorities have the capacity to devise plans to spend the finance and manage the projects as they come forward. Two decades of funding cuts have left many authorities with a much reduced staff resource. This is compounded by the fact that certain skills needed to respond to the new agenda do not necessarily exist in local government transport departments, and many are not often prevalent in transport consultancies either. A shortage of transport planners has been a key concern in the industry for many years and this is exacerbated by the demands of the *10-Year Plan*, particularly in the short and medium term. This problem was not helped by Labour's reluctance to see transport as a significant issue early in its first term of office and thus invest in the development of skills in this area.

Furthermore, the recent shifts in Labour's policy emphases meant that local authorities have found themselves faced with messages that are to varying degrees contradictory. *A New Deal for Transport* emphasized the management of existing infrastructure resources, yet the *10-Year Plan* and MMSs all allocate large sums to new infrastructure very much in a 'predict and provide' mode (Chapter 4). The justification is far from clear but appears partly based on time. It is argued that in the short-term such measures are needed before investment in alternatives and changes to planning policies can 'bite'. There is merit in this argument but many of the proposals themselves will not come on stream until the end of the *10-Year Plan* and it is likely that investment in promoting continued levels of car ownership and use will undermine efforts in expanding public transport, walking and cycling and reducing the need to travel.

There are also concerns that, in a hangover from the 1980s when levels of finance were heavily constrained, local authorities still perceive that investing in the creation of new policy strategies is barely worth the effort. This situation has now changed somewhat as more money for transport comes on stream, but advocates of greater freedom of action for local authorities argue that developing better strategies will require yet more decentralization and transfer of power. This could have the effect of reducing the transaction costs associated with central government's control over local authority transport spending and potentially create more money for projects themselves.

Financial issues also create a major stumbling block in the integration of transport and land-use policies. Since development decisions are heavily influenced by economic arguments, planning decisions are often based primarily on jobs and new businesses rather than on strong transport or land-use criteria. Local authorities are often too fearful of losing out on new development (and jobs) to impose too many conditions on development, particularly in the case of major prestige developments.[28] In many authorities, there is political pressure for development at (almost) any cost, particularly in less prosperous areas. This has strong links with the following discussion about political difficulties of implementing local transport policy.

Cultural

Cultural issues relate principally to the values, attitudes and mindsets of the key players in the transport policy system. In the UK, engineers dominate transport planning at the local level and, while such practitioners have many useful skills, many of the demands of Labour's policy thrust in relation to both transport and beyond – of partnership and participation and of maintaining and enhancing place qualities – are not necessarily those well represented or understood in some parts of practice.[29] The value systems of practitioners nurtured in a different cultural climate where particular expertise was valued more highly may thus represent something of a difficulty here. Such instances can occur from strategic to micro levels such as highway design and parking standards. Similarly, there are difficulties for planning departments in recruiting staff, to development control posts in particular, where an awareness of issues such as the importance of design, development location, neighbourhood permeability and the impacts of new development on transport conditions are essential if transport policy aims are to be met. This has a knock-on consequence for the training of transport and planning professionals. While a key issue here is often a shortage of students, there are questions about how far transport and planning courses are equipping professionals to respond to much of Labour's emphasis on accessibility for all and place quality.

Yet the largest cultural difficulty arguably remains among the travelling public. Efforts to implement local policies that challenge a culture of car ownership and use often receive a great deal of criticism from those affected and the local media. Many solutions (such as reducing road space, reducing traffic speeds and dropping highway design standards) are counterintuitive which means that establishing public support is often difficult. In addition, car travel is the only infrastructure that is not paid for at the point of use and it remains easy and cheap in most parts of the UK. Consequently there is limited consideration of transport issues in people's decisions over residential location, or where there is, it is connected to perceptions of travel time by car and subsequent time-space 'centrality' in relation to a series of everyday tasks.[30] This implies improved levels of dialogue in transport decision and policy-making processes to present the evidence to support policy choices and present likely outcomes. There is evidence that many urban local authorities, where such an approach is more pressing, have been making such efforts and these and the success of initiatives such as TravelSmart show the potential rewards.

Political

There are many political issues that provide difficulties for Labour in implementing its transport policy aspirations. Labour has been frustrated by the difficulties of getting infrastructure schemes through the planning process. Much of this relates to opposition from the public and local politicians. Such NIMBY (Not In My Back Yard) and NIMTOO (Not In My Term Of Office) concerns are long-standing issues. This opposition is frequently contradictory in terms of values and actions but is unlikely to go away, although there are signs that greater attention to the process of policy and decision making can help alleviate such difficulties.[31]

Such issues are made more problematic by clientelism on the part of local politicians. Councillors and members of parliament frequently (and quite understandably) use transport issues as a platform on which to be elected. This can, however, lead to continued support for proposals despite evidence that these may be contrary to local objectives and may also represent poor value for money. The use of 'technical' methods by local officers can help prioritise such decisions in many instances, as has the adoption of LTPs as strategic documents.[32] There is, however, often a failure to see the links between certain proposals, typically long hoped for 'pet' road schemes where degrees of ownership and invested political capital are apparent, and strategic policy. Such support often undermines strategic transport, planning and environmental policy objectives and entails a drift toward 'pragmatic multi-modalism' which is unlikely to work as it perpetuates a culture of car dependence while diverting resources away from alternatives (Chapter 1).[33] Parochialism at all spatial scales from nations and regions to wards promotes such demands as councillors, politicians, and quangos are judged by the resources flowing into their 'patch'.

A further highly significant political lobby comes from those promoting road-based solutions. Inputs that can highlight the importance of particular links on business competitiveness for example are an essential element to a policy process but they are often, by definition, characterized by narrow thinking with regard to ecological and social impacts and with little recourse to evidence on links between infrastructure investment and economic competitiveness. Such concerns have also in many areas held up charging schemes despite evidence that such schemes are likely to have positive long-term benefits.[34] Local competitiveness concerns are compounded in remoter parts of the UK by concerns over 'peripherality'. Such concerns manifest themselves in calls for improved links to 'core' markets (typically mainland Europe and south east England) despite evidence that in many cases jobs and business are more likely to be drawn out of peripheral areas than drawn in.[35]

Organizational

A wide range of organizational difficulties have hindered the delivery of Labour's transport agenda. Some of these relate to the 'hard infrastructure' of policy making (the design of policy mechanisms and the structures of policy systems), whereas others relate to the 'soft infrastructure' (the operation of those policy systems).[36] In terms of the hard infrastructure, while the LTP has overcome many of the difficulties associated with TPPs, they are still bidding documents. Funding for projects in LTPs as well as those devised in other arenas is still far from guaranteed over the longer term despite the existence of the *10-Year Plan*. This undermines the legitimacy of the LTP as a document.

The renewed importance of regional level planning in England and Scotland is potentially useful in transport planning as it may help to secure agreement over wide areas on more strategic priorities and avoid a degree of interlocality competition, enabling more radical policies to come forward. In the absence of elected regional government, these sorts of agreements have not been much in evidence so far and emerging RTSs tend to focus on 'regional-scale' projects, such as large new or improved pieces of infrastructure, at the expense of larger numbers of local initiatives that could have greater cumulative impact.[37]

In relation to soft infrastructure, the coordination of transport with land-use planning has been progressing well in many areas. At the regional level and in LTPs greater linkages appear to be being made with planning objectives and other policy areas. This ability to construct and maintain broad policy networks is another vital skill given greater importance in the Labour era as recognition grows of the dependence of meeting transport objectives, such as road traffic reduction, on other policy areas. Finally, outside of the public sector, there are also questions about the growth of monopoly and duopoly situations in local public transport markets. Whilst this can be useful in coordinating activity in areas such as ticketing and information provision it can also prevent the growth of local markets (Chapter 7). This becomes particularly problematic where relations between monopoly operators and local regulators such as PTEs have difficult histories.

These four sets of issues – financial, cultural, political and organizational
– combine to varying degrees in particular places to help and hinder the
implementation of Labour's objectives in relation to transport policy at the
local level. We now make some broad conclusions about Labour's approach
to local transport issues and look to the future for UK transport policy in
the light of this discussion.

Conclusions

There are many indications that the shift in Labour's first term from a
strong line on demand management towards a 'pragmatic multi-modalism'
will not help local authorities come up with policies that work in sustainable
transport terms.[38] Evidence suggests that a coherent package of measures is
what is needed at local levels within consistent regional and national policy
frameworks, both within transport and beyond. Given the need for this
coherence, transport policy making that remains characterized, at regional
scales for example, by local political horse trading is deeply inappropriate as
such processes are unlikely to arrive at effective packages. While government regional offices in England are policing many proposals to assess the
extent to which they might encourage road traffic growth, many larger
schemes appear not to be subject to strict demand management criteria.
The contradictory policy emphases in the *10-Year Plan* and *A New Deal for
Transport* do not help in this regard. The White Paper's attempts to position
transport's role as one of invigorating places, neighbourhoods, cities and
regions has been lost in this policy obfuscation. In addition, the emphasis
on congestion and time savings is unlikely to work and serves as a distrac-
tion.[39]

Without clear policy frameworks, more subtle issues such as conflicting
targets – for example, in the short-term increasing cycling may lead to more
road casualties – cannot be readily considered. The importance attached by
central government to Best Value indicators could make for unintended
consequences here as local authorities try and move forward on these
different fronts. There is also evidence that while Labour's emphasis on
public transport investment is working locally in some areas, such as light
rail schemes in Nottingham and South Hampshire (Chapter 6), other
issues, such as road safety, have not been taken up extensively by the
majority of local authorities despite central government exhortations.[40]

There are lingering issues concerning the 'soft infrastructure' of transport
policy making. Much of the Labour agenda requires new thinking and ways
of acting. Approaches to transport planning that look at the needs of places
first rather than highway capacities, that deal with the equity implications of
proposals first and foremost, and that crucially entail liaising extensively

with other policy communities both in policy and decision making, all provide somewhat different challenges than the transport planning practice of the 1980s. Such cultural and organizational change takes time and a consistent national policy framework would help in this regard. The *10-Year Plan* and the Transport Planning Skills Initiative are welcome innovations but the devil here is in the detail in both cases. It would be unfair to suggest that no progress has been made in bringing land-use planning and transport decision making closer together since 1997, but the two areas of policy are still not fully 'joined-up' (Table 3.3).

The policy consistency question also has implications for devolution and central government's role. National government's commitments to targets nationally and internationally implies a continued gatekeeper role for Westminster. That said, a more hands-off approach by the centre does mean that there will be more local variation in transport policy. There are questions over the local capacity to achieve Labour's aims – in some places the difficulties are financial, whereas in others eroded professional expertise and fossilized decision making structures provide more of a problem. These latter authorities need continued guidance and support, while those with more capacity may be able to deliver innovative and interesting policies in a way not experienced in the past if they are given the freedom and the tools to do so. New legislation such as that on Quality Partnerships and charging schemes contained in the Transport Act (2000) is an important enabling step, but needs to be backed up by further commitment and action from central government (Chapters 4 and 7). Nevertheless, the interdependent relationship of the local and the centre has been thrown into the spotlight in a way that it hasn't been for some considerable time.

Such policy-making efforts will be closely linked to the ability of localities to engage their citizens with the practice of transport planning. There can be problems of public acceptance of the more radical elements of 'new realist' transport policy although these are not insurmountable.[41] In many areas such measures have been implemented successfully. Here careful attention has been paid to implementation with an emphasis on dialogue with the public and other stakeholders, and on partnership between stakeholders within the transport policy network and beyond. In such places the public mood is, as *A New Deal for Transport* suggested in 1998, 'ripe for change'. Policies everywhere will need to be developed collaboratively if they are to succeed. There is evidence that a more holistic, place focused, 'joined-up' approach to transport planning can be made to work.[42] Transit systems and traffic calmed attractive neighbourhoods and central areas can and are being positioned as assets in competing for highly mobile workers and capital. While Labour's approach rightly recognizes that there is no 'one size fits all' transport strategy for localities, much is to be learnt from such areas. The challenge for other localities is to combine the growing

Table 3.3 Examples of integration and lack of integration between land-use planning and transport policy

Integration	Lack of integration
• Formation of the Department of the Environment Transport and the Regions (DETR) in 1997, which brought together government departments responsible for land use planning and transport policy.	• Creation of the Department for Transport (DfT) and the Office of the Deputy Prime Minister (ODPM) in 2002, which once again separates responsibility for transport and planning policy.
• Local government reorganisation, which created unitary authorities in previously two-tier authorities, has brought land use planning and transport decision-making 'under one roof'.	• Problems of spatial integration and coordination across new unitary authorities: areas covered by unitary authorities are small in relation to travel patterns.
• The LTP system is built around 5-year integrated transport strategies instead of an annual bidding round under the TPP system.	• Joint working on strategic planning issues are non-statutory.
• The *10 Year Plan*, was produced in 2000 with the intention of providing a more stable climate for investment in transport for both public and private sectors, given the typical investment horizons needed to plan for and deliver transport improvements.	• Politicians at all levels are more interested in policies that will have an effect sooner rather than later. • Political decision-making at the local level is often beset by parochial or NIMBY attitudes which work against strategic, joined-up decision-making.
• Transport assessments required under PPG13 summarise the transport implications of a planning application.	• Uncertainty about the ability of the private sector to maintain and expand infrastructure (for example, Railtrack).
• Planning policy guidance on transport (PPG13) gives advice on integrated land use planning and transport policy development.	• Transport assessments (and UDPs) do not identify cumulative impacts on national transport systems or knock-on effects for adjacent authorities.
• Planning policy guidance on regional planning (PPG11) requires the preparation of a Regional Transport Strategy which sets the long-term strategic framework for transport policies and proposals to be included in the development plan and the Local Transport Plan.	• Development decisions are heavily influenced by narrow economic arguments, often resulting in poor decisions based on parochial practices rather than strict location criteria.

- Policy guidance on development plans (PPG12) provides advice about maintaining consistency between local transport plans and development plans, including a section on joint working between agencies.

- Transport policy is often reactive to land use planning decisions: trying to address the consequences of earlier planning decisions made without much forethought for transport.
- Lack of real interdisciplinary team working between departments responsible for land use planning and transport policy.

evidence base in transport on the links between transport policy, economic competitiveness, ecological and social sustainability to continually confront the difficulties described earlier and drive appropriate change in their locales.

NOTES

1 Department of the Environment, Transport and the Regions (1998) *A new deal for transport: better for everyone.* Cmnd 3950, The Stationery Office, London.
2 The Regional Transport Strategy forms part of Regional Planning Guidance and provides the long-term strategic framework which informs development plans, local transport plans and transport operators in developing their plans and programmes. See Department of the Environment, Transport and the Regions (1999) *Planning policy guidance note 12: development plans.* The Stationery Office, London.
3 See Department of the Environment, Transport and the Regions (2000) *Guidance on full local transport plans.* DETR, London.
4 Commission of the European Communities (2001) *European transport policy for 2020: time to decide.* COM (2001) 370. Office for Official Publications of the European Communities, Luxembourg.
5 Cabinet Office (2000) *Wiring it up. Whitehall's management of cross-cutting policies and services.* The Stationery Office, London.
6 Stead, D and Banister, D (2001) Influencing mobility outside transport policy. *Innovation*, 14, 315–330.
7 Vigar, G (2002) *Participation and learning in the development of transport strategies: a case study of the North East of England.* Address to AESOP Congress, Volos, July.
8 Headicar, P (2002) Regional transport strategies: fond hope or serious planning? In Marshall, T; Glasson, J and Headicar, P (eds) *Contemporary issues in regional planning.* Ashgate, Aldershot; Vigar, G (2002) *Participation and learning in the development of transport strategies.*

9   Commission for Integrated Transport (2000). *Study of European best practice in the delivery of integrated transport. Report on stage 3: transferability.* CfIT, London.

10 Vigar, G; Healey, P; Hull, A and Davoudi, S (2000) *Spatial strategy and the English planning system: an institutionalist analysis.* Macmillan, London.

11 Department of the Environment and Department of Transport (1994) *Planning policy guidance note 13: transport.* HMSO, London; Ecotec (1993) *Reducing transport emissions through land-use planning.* HMSO, London.

12 Royal Commission on Environmental Pollution (1997) *Twentieth report. Transport and the environment: developments since 1994.* The Stationery Office, London, 8.7.

13 Gorham, R (1998) *Overcoming barriers to effective co-ordination.* Background document for the OECD-ECMT workshop, Land-use for sustainable urban transport: implementing change, Linz, 23–4 September. OECD-ECMT, Paris.

14 Stead, D (2003) Transport and land-use planning policy in Britain – really joined up? *International Social Science Journal,* in press.

15 Allmendinger, P (2002) *Integrating planning in a devolved Scotland.* Paper presented at the Planning Research Conference, University of Dundee, April.

16 Vigar, G; Steele, M; Healey, P; Nelson, J and Wenban Smith, A (2000) *Transport planning and metropolitan governance.* Landor, London.

17 Begg, D and Docherty, I (2002) *The future of strategic transport planning in Scotland: rediscovering the city-regional approach.* Policy Paper Series No. 9, The Centre for Transport Policy, Aberdeen.

18 Royal Commission on Environmental Pollution (1997) *Twentieth report,* 7.28.

19 The absence of any significant consideration of the land-use, and wider social consequences, of the *10–Year Plan* should also be noted. See Hall, P and Marshall, S (2002) *The land-use effects of the 10–Year Plan.* Independent Transport Commission, London.

20 Department for Transport (2002) *The government's response to the Transport, Local Government and the Regions Select Committee Report, 10–Year Plan for Transport.* Cmnd 5569. http://www.dft.gov.uk/trans2010/response/pdf/response.pdf (accessed 3 November).

21 The Single Capital Pot is, as its name suggests, an arrangement whereby councils are free to make their own decisions about the allocation of capital funds between different services.

22 See Docherty, I and Hall, D (1999) Which travel choices for Scotland? A response to the government's white paper on integrated transport in Scotland. *Scottish Geographical Journal,* 115, 193–210.

23 Best Value is the requirement that (local) government policies are benchmarked in terms of their value for money.

24 Department of the Environment, Transport and the Regions (1998) *A new deal for transport,* 18.

25 Bickerstaff, K and Walker, G (2001) Participatory local governance and transport planning. *Environment and Planning A,* 33, 431–51.

26 Vigar, G (2002) *Participation and learning in the development of transport strategies.*

27 Vigar, G (2000) Local 'barriers' to environmentally sustainable transport planning. *Local Environment*, 5, 21–34; Vigar, G; Steele, M; Healey, P; Nelson, J and Wenban Smith, A (2000) *Transport planning and metropolitan governance*.

28 Royal Commission on Environmental Pollution (1997) *op cit.*, 7.26.

29 Vigar, G (2000) Local 'barriers' to environmentally sustainable transport planning; Vigar, G; Steele, M; Healey, P; Nelson, J and Wenban Smith, A (2000) *Transport planning and metropolitan governance*; Bickerstaff, K and Walker, G (2001) Participatory local governance and transport planning.

30 Champion, T and Ford, T (2001) *Who moves where and why? A survey of residents' past migration and current intentions. Results of the Newcastle case study.* Report for Newcastle City Council, Department of Geography, University of Newcastle; Jarvis, H; Pratt, A and Wu, P (2001) *The secret life of cities: the social reproduction of everyday life.* Pearson Education, New York.

31 Blake, J (1999) Overcoming the value-action gap in environmental policy: tensions between national policy and local experience. *Local Environment*, 4, 257–78; Burningham, K and O'Brien, M (1994) Global environmental values and local contexts of action. *Sociology*, 28, 913– 32; Vigar, G; Steele, M; Healey, P; Nelson, J and Wenban Smith, A (2000) *Transport planning and metropolitan governance*; Vigar, G (2002) *The politics of mobility: transport, the environment and public policy.* Spon, London.

32 See Docherty, I (2000) Rail transport policy-making in UK Passenger Transport Authority areas. *Journal of Transport Geography*, 8, 157–70.

33 Shaw, J and Walton, W (2001) Labour's trunk-roads policy for England: an emerging *pragmatic multimodalism. Environment and Planning A*, 33, 1031–56.

34 Whitehead, T; Preston, J and Holvad, T (2002) *The whole-life impacts of transport charging interventions on business performance: a time-marching framework.* Transport Studies Unit, Oxford, Ref. 921.

35 Chandra, A and Thompson, E (2000) Does public infrastructure affect economic activity? Evidence from the rural interstate highway system. *Regional Science and Urban Economics* 30, 457–90; Whitelegg, J (1997) *Critical mass.* Pluto Press, London.

36 Healey, P (1997) *Collaborative planning.* Macmillan, London.

37 Porter G and Vigar, G (2003) Governance, accountability and responsibility in the construction a regional transport strategy for the North East. *Northern Economic Review*, in press; Headicar (2002) Regional transport strategies.

38 Shaw, J and Walton, W (2001) Labour's trunk-roads policy for England.

39 Goodwin, P (2001) *Running to stand still? An analysis of the ten year plan for transport.* Council for the Protection of Rural England, London.

40 House of Commons (2002) *Road traffic speed.* Ninth report of the Select Committee on Transport, Local Government and the Regions. Session 2001–2002, HC 557–I, 13 June, The Stationery Office, London.

41 Cairns, S; Atkins, S and Goodwin, P (2002) Disappearing traffic? The story so far. *Municipal Engineer*, 151, 13–22.

42 Jacobs, M (1999) *Environmental modernisation, the new Labour agenda.* The
    Fabian Society, London; Vigar, G; Steele, M; Healey, P; Nelson, J and Wenban
    Smith, A (2000) *Transport planning and metropolitan governance.*

Part II

Progress in Policy Implementation

4

Roads and Traffic Congestion Policies: One Step Forward, Two Steps Back

William Walton

I will have failed if in five years' time there are not many more people using public transport and far fewer journeys by car. It's a tall order but I urge you to hold me to it.[1]

This country is quite small . . . The transport solutions of the future cannot be to pour concrete over large parts of England's green and pleasant land.[2]

A central objective – indeed arguably *the* central objective – of Labour's 1998 transport White Paper, *A New Deal for Transport: Better for Everyone*, was to tackle rising traffic congestion and associated pollution.[3] This was to be achieved through the implementation of 'carrot and stick' policies that would result in 'persuading people to use their cars a little less – and public transport a little more' (Chapter 1).[4] The carrot consisted of promises to significantly improve the provision of public transport through increased levels of investment to encourage people to forsake their car and opt instead for the bus, train, light rail and so on. Those drivers deciding not to do this would suffer the consequences of the punitive stick policies in the combined form of vastly reduced expenditure on road construction – irrespective of any growth in traffic levels – and the imposition of charges for road access and workplace parking.

Labour's new, more restrictive approach to car usage was postulated as representing something of a reverse from the permissive policy of 'predict and provide' followed by the Conservatives during the larger part of their 18 years in power. Predict and provide, which reached its zenith with the publication in 1989 of the White Paper *Roads for Prosperity*, was presented as an attempt by ministers to build sufficient new roads to accommodate anticipated levels of traffic growth.[5] This was considered necessary to facilitate the free flow of traffic that the government viewed as being desirable for economic, safety and ideological reasons (it regarded car ownership as being symbolic of its proclaimed individualist ethos). Importantly, notwithstanding the Conservatives' predisposition to the privatization of publicly owned utilities, there was never any serious suggestion of seeking to resolve the problem of growing traffic congestion through the introduction of a market-based mechanism that would require road users to pay access charges (that is, road user charging). Indeed, since the Conservative government conceptualized congestion as being no more than an unfortunate but unavoidable by-product of economic growth, there was a clear implication that any measure that might be introduced to 'manage' demand for road space might only serve to jeopardize future prosperity. Most concerns about resultant levels of pollution were equally brushed aside on the grounds that developments in fuel and engine technology would cancel out any deleterious environmental impacts caused by rising levels of car usage.

By the early 1990s this transport policy began to unravel as a result of an increasingly unreliable bus and railway network, and the growing perception that the seemingly never ending programme of new road construction was unlikely to resolve the spiralling problems of traffic congestion. Labour was quick to recognize the political opportunity presented by the disaffection generated by these problems, in particular those associated with road building which regularly attracted vociferous and hostile public opposition. High-profile demonstrations were mounted against the M65 south of Blackburn, the M11 through east London and, perhaps most famously, the A34 west of Newbury. Although many of the demonstrators, such as the now legendary 'Swampy', were self-styled 'new age' ecological warriors who advocated lifestyles and values very different to the mainstream of society, their opposition to mass road building appeared to chime with the views of an increasing number of people from 'middle England'.[6] This opposition gained support in 1994 from the publication of two independent and highly authoritative reports by the Royal Commission on Environmental Pollution (RCEP) and the Standing Advisory Committee on Trunk Road Assessment (SACTRA).[7] Both cast serious doubt on the wisdom of the Conservatives' predisposition to accommodating the needs of the private car.

Throughout the 1990s, Labour sought to drive transport issues further up the political agenda and highlight its own alternative. In its policy statement issued shortly before the 1997 general election, *Consensus for Change*, the Party argued that there was a broad agreement across a wide range of interests that trends in car usage were unsustainable.[8] In response to this consensus, Labour sought to shape a policy framework that would 'keep the country moving and breathing . . . ' by 'moving beyond predict & provide', towards a genuine improvement in public transport provision. As part of its manifesto commitment, Labour pledged to introduce a wide-ranging review of the national road building programme and, following its election victory in May 1997, began to construct a framework within which it could implement its plans. This pledge was incorporated into *A New Deal for Transport* alongside proposals for road user charging and a workplace parking levy as part of a wider plan to effect modal shift.

This chapter examines Labour's roads and traffic policies since its election to government in 1997. It analyses the level to which measures to reduce road building and promote road user and workplace parking charges have been implemented at the national and local level. Although there are references to the whole of the UK, the focus of this chapter is very much on the application of Labour's policies, as set out in the White Paper and *Transport 2010: The 10-Year Plan*, to England.[9] This is essentially because, in policy-making terms, England is some way ahead of Scotland, Wales and Northern Ireland, whose devolved administrations are now advancing their own transport strategies in corresponding documents (inspection of these documents reveals approaches broadly similar to that adopted in England) (Chapter 2).[10] Moreover, because the policy programme is further advanced in England it has already been subject to some detailed scrutiny and this provides the basis for some of the figures contained in this chapter.[11] But before considering whether Labour's policies have begun to succeed it is necessary to review the nature of the roads policies pursued by the Conservatives between 1979 and 1997.

Roads and Traffic Policy under the Conservatives

At the time of the Conservative Party's election to government in 1979 there were around 14 million cars on Britain's roads. These were accommodated on a principal road network that had largely been constructed during the post-war period which, although almost 'complete' in comparison with that in place today, still lacked a number of important sections, such as the M25 around London and the M40 between Oxford and Birmingham. In order to accommodate further anticipated increases in car ownership and usage, the Conservatives, in line with previous post-war governments, planned to expand this network by filling in the 'gaps'. Following public spending reviews, firm commitments to new road schemes, which would usually take somewhere between 9 and 13 years to progress from the drawing board to completion, were identified for the immediate 12-month period. Schemes beyond this time horizon were identified only provisionally subject to the availability of future finances. Between 1979–97 there was an 'active' rolling trunk-road building programme of between 850–1000 miles in length with a reserve 'long term' programme of between 180 to 610 miles (Table 4.1). Levels of completions

Table 4.1 Summary of the Conservatives' trunk-road building programmes, 1979–97

Publication	Period covered	Active programme	Long-term programme	Completions during period	Average annual completions during period	Average completed scheme length during period
Policy for roads: England 1980. Cmnd 7908	Feb 1978-Jan 1980	179 schemes 890 miles	166 schemes 609 miles	47 schemes 170 miles	23.5 schemes 85 miles	3.62 miles
Policy for roads: England 1981. Cmnd 8496	Feb 1980-Dec 1981	180 schemes 850 miles	78 schemes 219 miles	44 schemes 153 miles	21 schemes 73 miles	3.47 miles
Policy for roads in England: 1983. Cmnd 9059	Jan 1982-June 1983	218 schemes 944 miles	64 schemes 182 miles	32 schemes 96 miles	24 schemes 64 miles	2.7 miles
National roads: England 1985	July 1983-May 1985	242 schemes 931 miles	92 schemes 359 miles	52 schemes 168 miles	25 schemes 81 miles	3.24 miles
Policy roads in England: 1987. Cmnd 125	June 1985-Apr 1987	234 miles 978 miles	120 schemes 515 miles	50 schemes 180 miles	24 schemes 86.5 miles	3.6 miles
Roads for prosperity (1989). Cmnd 693	May 1979-Dec 1989[1]	494 schemes 2800 miles	N/A[2]	264 schemes 900 miles[3]	26 schemes 90 miles	3.46 miles
Trunk roads in England: 1994 review	Jan 1990-Jan 1994	256 schemes 1341 miles	67 schemes	128 schemes 476 miles	42.7 schemes 158.5 miles	3.71 miles
Managing the trunk road programme (1995)	Feb 1994-Nov 1995	138 schemes 853.3 miles[4]	104 schemes	34 schemes 126 miles	15.6 miles 59 miles	3.78 miles

1996 trunk road review. DoT press notice 357	Dec 1996–Nov 1996	132 schemes 573 miles[5]	0 schemes[6]	16 schemes 63 miles	16 schemes 63.3 miles	3.96 miles
Total	Feb 1978–Nov 1996	489 schemes 1735 miles		25.7 schemes 91.3 miles	3.55 miles

Source: Shaw, J and Walton, W (2001) Labour's trunk road building proposals for England: an emerging pragmatic multimodalism? *Environment and Planning A*, 33, 1031–56.

NOTES

1 *Roads for Prosperity* did not give 1987–9 completion figures, providing instead data on completions achieved during the first 10 years of Conservative rule

2 *Roads for Prosperity* did not distinguish between the active and long-term programmes

3 Total number and length of schemes completed 1979–1989

4 Excludes mileage from 26 Design, Build Finance and Operate (DBFO) projects as this was not included in the White Paper

5 Excludes mileage from DBFO projects

6 The long-term list was withdrawn in 1996

averaged between 64 and 86.5 miles per year.[12] The government's main priority during this period was for the completion of the motorway orbital routes around London (M25), south Birmingham (M42) and Manchester (M60). The need for new bypasses around historic towns (such as Colchester) was also recognized as important, even though, due to lack of funds, this meant delaying progress on other strategic schemes such as the extension of the M20 and M65 in Kent and Lancashire respectively.

For a government committed to a 'great car economy', the massive increase in the road building programme announced in the 1989 White Paper, *Roads for Prosperity*, represented a logical response to the problems of congestion resulting from increasing levels of car ownership and usage. This policy was announced by Paul Channon, then Secretary of State for Transport, in his speech to the House of Commons:

> We have reviewed traffic trends and produced new forecasts. Those are set out in the White Paper. They project substantially faster growth than previously estimated. In the light of the new forecasts, the Government have concluded both that our main efforts to provide additional transport capacity in support of growth and prosperity must be directed towards widening existing roads and building new ones, and that a step change in the programme is essential . . . Accordingly, I can inform the House that the road programme will be more than doubled, from over £5 billion to over £12 billion.[13]

Roads for Prosperity reported that the 'more-than-doubling' of road building expenditure was to manifest itself through the construction of over 1,450 miles of new and widened trunk roads. Although most of the new roads identified in the 1989 programme were to be paid for through the public purse, a green paper issued simultaneously anticipated the need for future involvement of the private sector in the promotion, finance and operation of new trunk roads.[14] It dismissed the notion of using 'shadow tolls' – payments to the private sector based on levels of vehicle usage for operating and managing new and existing roads – preferring instead the direct involvement of the private sector in promoting new toll roads. The Western Orbital Road and the Birmingham Northern Relief Road (BNRR), motorway standard roads in the Midlands designed to relieve the congested M5 and M6, would be the first to be put out to private tender (plans to build the Western Orbital Road have subsequently been abandoned).

Whilst the government had foreseen that the programme might be difficult to fund in full, it had probably not anticipated how quickly the fortunes of the public sector finances would change and also how controversial individual road schemes would prove to be. As a result of these factors the generously expanded roads programme soon became a target for criticism from a wide range of interests.[15] It was being argued in some quarters that instead of unconditionally attempting to accommodate rising levels of car usage, the government should introduce measures to manage levels of demand and promote public transport.[16] Strong technical support for this 'new realism' was provided by the publication of the Eighteenth Report of the RCEP in 1994 and by the SACTRA report of the same year on induced traffic. The RCEP implored the need to reduce the level of individuals' car dependence arguing that it was fundamentally unsustainable. Amongst its 110 recommendations was the need to halve the scale of the road building programme and double the real price of fuel by 2005. SACTRA, the government's own advisory body on trunk-roads assessment, reported its findings on the little understood concept of induced traffic (which up to then had been labelled the 'M25 effect'). These demonstrated that under certain conditions the provision of new road space could induce car owners to make journeys they otherwise would not have made. As a result the technical justification for many new road schemes, which had until then been evaluated on the assumption that the level of car usage was independent of the amount of road space (the fixed trip matrix assumption), was seriously weakened.

The immediate and largely predictable public hostility to a suggestion of motorway tolling to pay for new roads meant that the idea was quietly dropped, leaving ministers faced with the realities of a tightened budget little choice but to reduce the size of the national road programme.[17] The main problem was that the schemes identified in the 1989 programme could not be progressed at equal speed and there was therefore a need to produce a rationalized and more clearly focused programme that had a greater prospect of delivery. Whilst the government continued to argue that new roads were important to the overall health of the economy, the tone of its claims were considerably more muted. Implicitly, there was an acceptance of the occurrence of induced traffic – 'We cannot deal with the problems of increasing traffic simply by road building . . . ' – and explicitly there was a recognition of the growing imperative of sustainable development, through the need to ' . . . strike the right balance between securing economic development, protecting the environment and sustaining future quality of life'.[18]

The implicit rejection of predict and provide represented an important watershed in transport policy.[19] From 1994 through to their departure from office in 1997, the Conservatives continued to cut the size of the trunk-road building programme and, in tandem, there was sharp fall in the number of local bypass completions from 1995/96 onwards (Table 4.2). Initially the trunk-road programme was reduced from around 500 schemes to around 320, returning it to a scale more in line with those announced in the early 1980s. In 1995, in the wake of the national 'Transport Debate' initiated by the then Secretary of State for Transport, Brian Mawhinney, a further series of cuts was made and a year later another 77 schemes

Table 4.2 Major local road schemes completed 1985/86–1997/98

	85/6	86/7	87/8	88/9	89/90	90/1	91/2	92/3	93/4	94/5	95/6	96/7[2]	97/8[2]	85–98 total completions	85–98 mean annual completions
Bypasses	26	23	25	28	24	23	20	21	20	20	17	11	11	269	20.7
Relief roads	26	22	22	17	28	19	22	11	12	13	16	22	12	242	18.6
Other	21	18	23	19	15	14	14	8	10	11	8	10	12	183	14.1
PRN[1]	46	42	50	50	48	40	47	16	21	30	27	27	22	466	35.8
Total Km	N/A	N/A	N/A	117	142	112	83	89	127	205	155	93	135	1,258[3]	125.8[3]
														(786 miles)	(78.6 miles)

Source: Walton, W and Shaw, J (2003) Applying the new appraisal approach to transport policy at the local level in the UK. *Journal of Transport Geography*, 11, 1–12.

NOTES

1   Aggregate of bypasses, relief roads & 'other' forming part of Primary Road Network
2   Projections
3   Figures for 1988–98 only: no aggregate figures were available prior to 1988/89

were removed. By the time the Conservatives left office in 1997, just 146 schemes remained in the trunk-road building programme, and road building per se had been largely discredited as a policy response to what was by then seen as a chronic national transportation problem. Predict and provide was, to all intents and purposes, a thing of the past.

During their years in office the Conservatives had presided over the construction of around 490 trunk-road schemes totalling around 1,735 miles, equivalent to the construction of about 95 miles per year. The great majority of this additional road space was in the form of dual carriageways, but a small number of new motorways was included, some of which were environmentally controversial (for example, the extension of the M3 over Twyford Down).[20] This expansion of the national trunk-road network was paralleled by an expansion of the local road network. A preliminary analysis of local road building 1985/86 to 1997/98 in England shows that a total of 511 bypasses and local relief roads was constructed (with a further 183 lesser roads classified as 'other'), of which 466 formed part of the Primary Route Network.[21] Between 1988/89 and 1997/98, a total of 786 miles of local roads was built, equivalent to around 79 miles per annum.[22]

The New Approach to Roads and Traffic

Labour's concerns about the dominance of the private car are easy to comprehend. By 1997 levels of car ownership had risen to around 20 million and usage was also on a continuing upward trend. Social benefits resulting from increased levels of mobility and accessibility are being severely undermined and, arguably, outweighed by a host of associated problems (Chapter 1). Road building has led to the irreversible loss of some of the UK's most treasured open countryside.[23] High volumes of car usage has meant that the quality of air in many urban areas now fails to comply with standards laid down in the Conservative government's *United Kingdom National Air Quality Strategy*, potentially hastening the deaths of around 24,000 people each year.[24] Many parts of the road network are now operating well beyond their design and operational capacity, with the position expected to deteriorate significantly further by 2016. Increasing car dependency also produces adverse social impacts. Those without access to private vehicles lose out as the availability of public transport and other services such as shops declines in the face of falling demand, thus creating difficulties of mobility and accessibility.

Labour's proposals for roads and traffic in England are set out in detail in *A New Deal for Trunk Roads in England*, a 'daughter' document of *A New Deal for Transport*. The prioritization of the funding for these proposals is set out in *Transport 2010*. Labour's first principal objective is to reduce traffic congestion on interurban trunk roads to below 2000 levels by 2010. Although traffic congestion is a clearly understood concept, defining and measuring it turns out to be more problematic. The government defines it as the amount of time lost over a section of highway through the need to reduce speed below the legal limit (presumably assuming good driving conditions) due to other vehicles. As Phil Goodwin has suggested, whilst the use of this measure might appear perfectly proper and legitimate, on application it can be shown to produce spurious results which are difficult to interpret and monitor.[25] Progress towards the second principal objective of emissions reduction was to be measured through changes to ambient pollution levels recorded at roadside monitoring stations sited across the UK. Here, again, there is some scope to achieve apparent compliance by creatively interpreting the wording in the directive governing the siting of air monitoring stations, although this dubious practice appears to be giving way to a more *bona fide* approach.

Road Building under Labour

Prior to the publication of *A New Deal for Transport*, it appeared that significant cuts in road building were central components of Labour's declared transport strategy. Within months of coming to power in 1997, Gavin Strang, the Minister of State for Transport, launched the government's promised strategic review of the national trunk-road building programme and gave a commitment that new roads would only be built as 'a last resort'.[26] A moratorium was placed on all outstanding schemes, except for those at an advanced stage of construction and 12 proposals considered sufficiently urgent to be placed in an accelerated review.

This review was conducted under the New Approach to Appraisal (NATA), designed to provide a more sophisticated and rigorous assessment of new road proposals than that which the Conservatives had used. The way in which user benefits were defined and measured under the previous appraisal technique invariably favoured new roads over other transport proposals. Under the system of cost benefit analysis (CoBA) used to appraise and prioritize competing road schemes, 85 per cent of the appraisal weighting was given to projected time savings for drivers with the remaining 15 per cent of benefits being accounted for by reduced vehicle operating costs and improved safety. Any road proposal that, when discounted over a period of time, produced a net-positive cash return (that is, where the anticipated value of benefits outweighed that of costs) would be supported in principle subject to environmental considerations. These were examined through a separate system of Environmental Assessment (EA) in which impacts were expressed in nonmonetary terms. Because detailed environmental considerations were only undertaken for those schemes with a positive CoBA score, there was a sense that their importance was very much secondary to economic benefits, in particular those relating to anticipated time savings for drivers. As a result, it was these considerations that often appeared to provide the momentum for the road building programme.

Labour ministers regarded reform of the appraisal mechanism an essential component of their move towards a more sustainable transport strategy as it would be designed to make the promotion of new roads at both the national and local levels more difficult. Under NATA, cost benefit techniques, formerly restricted to *intra*modal investment evaluation, would now be used to appraise the relative benefits of competing *inter*modal investment proposals (for example, a new road versus a new rail line). In addition, two new criteria – accessibility and integration – were added to the existing three of safety, economy and environmental impact to reflect the new thinking on transport provision. The results of the accelerated review were announced within a month. Five (including the BNRR) of the 12 reviewed schemes would proceed, and the remaining seven were allowed to progress through their preparatory stages so that if, after full appraisal at a later date, they were to be selected they could be taken forward without further delay.[27]

The government's attitude to road building appeared to change at some point following the accelerated review. In the run-up to the release of the *10-Year Plan* a number of stories circulated in the press suggesting that Labour's commitment to reducing the scale of the roads programme was about to end and that a very large expansion was planned. These rumours were confirmed when the *Plan* earmarked a considerable amount of money for investment in trunk and local road construction. Around £5.4 billion – almost all from the public sector – will be invested in strategic roads before the end of 2003/04, representing about 23 per cent of all investment in transport during that period. The total level of planned public and private expenditure on roads over the 10-year period is around £59 billion, around one third of the total planned expenditure of £180 billion and 41 per cent of the total of planned public funded investment. £21 billion will be spent on trunk roads. The principal road building elements of the *10-Year Plan* are set out below.

The Targeted Programme for Improvements

The Targeted Programme for Improvements (or 'Targeted Programme') consists of those trunk road schemes to which the government is committed to building and was announced in 1998. It initially contained 42 schemes but has been expanded to 58 following the addition of a further eight schemes in March 2001, two in August 2002 and six in December 2002 (Table 4.3).[28] This programme is significantly smaller than that inherited from the Conservatives, but sufficiently large to indicate the intention to tackle the worst bottlenecks of the road network through road building. Announcing the targeted programme to Parliament John Reid, the Minister, stated:

> Every year, they [the previous government] produced a fantasy football league of hundreds of road schemes . . . We are offering a policy programme based on a rational analysis which is practical, and which is funded . . . and which can be delivered within a specified time.[29]

To this effect the *10-Year Plan* reaffirms the government's commitment to implement the programme by 2010. The total length of the 58 schemes is approximately 323 miles, equivalent to the construction of around 32.3 miles per year over the plan period. Most of the schemes are new roads – such as the A6 Clapham bypass in Bedfordshire – but about a third, such as the 4.1 mile long A3 Hindhead road in Surrey, are defined as 'improvements' in the form of widening designed to enhance capacity. Progress on implementation has been swift. By the end of 2002, around two-thirds of the schemes were either completed, under construction or had had contracts awarded.[30] The government anticipates that all of the schemes within the Targeted Programme will be completed or well under way by 2008. To fund this, the investment in road building from 2000/01 to 2006/07 is being increased from £0.9bn to around £1.5bn per annum, and then again to around £1.7bn per annum 2007/08 to 2010/11. Such a large-scale commitment to trunk-road building is being mirrored in Scotland, Wales and Northern Ireland where ministers have also announced major upgrade programmes.[31]

Trunk road bypasses and motorway widening

The *10-Year Plan* also provides for the construction of up to a further 30 trunk-road schemes from a list of 44 over and above those in the Targeted Programme, and the widening of around 360 miles of the national motorway network. The decisions on which bypasses and motorways are to be constructed or widened will be made within the context of multi-modal studies (MMSs) (Table 4.4).[32] The context for this new form of assessment was first mooted by the *United Kingdom National Air Quality Strategy*.[33] The document encouraged local authorities in England and Wales to take 'a strategic and multi-modal approach to addressing transport problems by bidding for funding for a package of transport measures rather than for discrete projects'.[34] This approach was developed in *A New Deal for Trunk Roads in England*, which stated that studies using NATA of 'the most urgent transport problems', in 21 (now 22) defined areas would be commissioned. Some would be 'road based, aimed at addressing particular localized problems on the trunk-road network, others . . . [would] be more wide ranging studies of transport problems across all modes of transport within certain key corridors and areas'.[35] Notwithstanding the publication of best practice guidance there has been some delay in the preparation of the studies, which is perhaps not surprising given the inherent complexity and relative novelty of attempting to model levels of intermodal transfer between say, car and bus over such large geographical areas.[36]

The government's initial expectation was that the predominant outcome of the MMSs – especially the larger, wide-ranging ones – would be non-road-based solutions. According to the guidance, 'there should be a strong presumption against [providing more roads] unless all other options can clearly be shown to be impractical', implying that very few of the 44 trunk-road schemes being reconsidered would be recommended for construction.[37] But given that the *10-Year Plan* subsequently gave a clear statement of intent to fund the construction of up to 30 of the trunk roads currently being evaluated, it appears that the government considers building new roads to be the easiest, or most desirable, option in many cases. If the *10-Year Plan*'s pledge is honoured, it could lead to the construction / widening of around 180 miles – or around 18 miles per year over the plan period – of trunk roads (about two-thirds of the total length of the 44 schemes). Those new roads approved will be incorporated into the Targeted Programme.

By the end of 2002, 15 of the 22 MMSs had reported, and the Secretary of State for Transport had made final decisions on eight. Table 4.5 shows that the construction and improvement of roads are important components in most of the strategies so far released. Although the first ministerial decision on an MMS, the *Access to Hastings Study*, resulted in the rejection of plans for two bypasses across scenic chalk downland, substantial road building schemes have been approved in the *Cambridge / Huntingdon Multi-Modal Study* and in the *South East Manchester Multi-Modal Study*.[38] In the former, the widening of the A14 and other minor highway improvements accounted for 74 per cent of the cost of a £261 million transport improvement package. In the latter, road spending only accounted for 20 per cent of the spending but this was still enough to fund the construction of the Alderley Edge, Poynton and Stockport bypasses and the A555 Manchester Airport Link Road.[39]

An illustration of the scale of the road schemes being recommended in some of the larger studies is provided by the *North / South Movements on the M1 Corridor in the East Midlands Multi-Modal Study* covering the area

Table 4.3 The Targeted Programme for Improvements

Scheme	Length (miles)	Scheme	Length (miles)
A1 Peterborough–Blyth	86.0	A43 M40–B4031 improvement	4.1
A1 Willowburn–Denwick	2.6	A46 Newark–Lincoln improvement	8.1
A1 (M) Ferrybridge–Hook Moor	10.2	A46 Newark–Widmerpool improvement	18.0
A1 (M) Wetherby–Walshford	3.7	A63 Selby bypass	6.1
A2 Bean–Cobham widening (Phase 1)	3.7	A66 Greta Bridge to Stephen Bank improvement	3.0
A2 Bean–Cobham widening (Phase 2)	2.5	A66 Carkin Moor to Scotch Corner improvement	3.8
A2/A282 Dartford improvement	2.1	A66 Stainburn & Great Clifton bypass	2.4
A2/M2 Cobham widening	9.0	A69 Haydon Bridge bypass	2.0
A3 Hindhead improvement	4.1	A120 Stansted–Braintree improvement	15.0
A5 Nesscliffe bypass	2.8	A249 Iwade–Queensborough improvement	3.3
A6 Great Glen bypass	3.1	A303 Stonehenge	5.7
A6 Rushden & Higham Ferrers bypass	3.5	A419 Commonhead junction improvement	4.7
A6 Rothwell–Desborough bypass	3.4	A421 Great Byford bypass	1.0
A6 Clapham bypass	3.7	A500 City Road & Stoke Road improvement	2.5
A6 Alvaston	1.4	A500 Basford/Hough/Sharington bypass	0.5
A10 Wadesmill Colliers End	6.7	A564 Derby southern bypass	2.0
A11 Roundham Heath–Attleborough improvement	6.2	A595 Parton–Lilyhall improvement	3.1
A13 Ironbridge–Canning Town	0.7	A650 Bingley relief road	4.2
A14 Haughey New St–Stowmarket improvement	3.0	A1033 Heden Road improvement	27.5
A21 Lamberhurst bypass	1.9	BNRR	
A23 Coulsdon inner relief road	1.1	M4 J18 eastbound diverge	1.3
A27 Polegate bypass	1.8	M5 J17–18a northbound climbing lane	1.3
A30 Bodmin–Indian Queens improvement	6.0	M5 J19–20 southbound climbing lane	1.3
A34 Chieveley/M4 J13 improvement	1.3	M5 J19–20 northbound climbing lane	1.3

A38 Dobwalls bypass	2.0	M25 J12–15 widening	6.8
A41 Aston Clinton bypass	4.0	M60 J5–8 widening	4.6
A43 Silverstone bypass	5.0	M62 Junction 6 improvement	5.0
A43 Whitfield Turn–Brackley Hatch improvement	2.8	M66 Denton–Middleton	5.6

Source: After Shaw, J and Walton, W (2001) Labour's trunk road building proposals for England: an emerging pragmatic multimodalism? *Environment and Planning A*, 33, 1031–1056.

Table 4.4 Schemes being considered in multi-modal studies

Scheme	Length (miles)	Multi-modal study
A1 Gateshead W bypass	6.7	Tyneside area study
A1 (M) J6–J8 widening	9.0	London–South Midlands study
A1 (M) Redhouse–Ferry	9.1	South and West Yorkshire motorway box study
A2 Lydden–Dover	6.0	RPC
A3 Hindhead	4.1	RPC
A5 Dunstable Eastern bypass	5.6	London to South Midlands study
A14 M11 J14–M1/A1 link	14.5	Cambridge to Huntingdon study
A21 Tonbridge bypass	2.2	Access to Hastings study
A27 Arundel bypass	3.4	So'ton to Folkstone study
A27 Wilmington bypass	1.0	So'ton to Folkstone study
A27 Selmeston bypass	1.0	So'ton to Folkstone study
A30/A303 Marsh Honiton	8.1	London to South West study
A38 Saltash improvement	1.7	London to South West study
A47 Thorney bypass	2.7	RPC
A63 Castle St Hull	1.0	Humberside study
A259 Bexhill/Hastings E bypass	3.8	Access to Hastings study
A259 Hastings W bypass	9.4	Access to Hastings study
A259 Pevensey	2.5	Access to Hastings study
A303 Ilminster bypass	6.4	London to South West study
A303 Ilminster Marsh	6.3	London to South West study
A303 Sparkford	3.4	London to South West study
A303 Wyle–Stockton Wood	2.5	London to South West study
A303 Chicklade bottom	8.0	London to South West study
A453 Clifton–M1	4.2	A453 M1–Nottingham study
A453 Clifton Lane	3.4	A453 M1–Nottingham study

A550 Deeside Park	3.6	Deeside Park junctions study
A556 (M) M6-M56 link	6.5	West Midlands to North West conurbation study
A590 High/low Newton bypass	2.4	RPC
A5117 M56	5.0	Deeside Park junctions study
M1 J6A–10 widening	14.0	London to South Midlands study
M1 J21A–23A widening	2.8	North to South movements in East Midlands study
M1 J23A–25 widening	7.0	North to South movements in East Midlands study
M1 J25–28 improvements	15.0	North to South movements in East Midlands study
M1 J31–32 widening	1.5	North to South movements in East Midlands study
M4 J4B–8/9 widening	14.0	London to Reading study
M6 Carlisle–Guardsmill extension	6.0	RPC
M6 J11A–16 widening	34.4	West Midlands to North West conurbation study
M6 J16–19 widening	16.3	West Midlands to North West conurbation study
M11 J14 improvement	3.4	Cambridge to Huntington study
M42 J3A–7 widening	10.3	West Midlands area study
M62 east–M606 link	0.4	RPC

Source: After Shaw, J and Walton, W (2001) Labour's trunk road building proposals for England: an emerging pragmatic multimodalism? *Environment and Planning A*, 33, 1031–56.

Table 4.5 Major road building elements in 'reported' multi-modal studies, as at December 2002

Multi-modal study	Final report	Principal road building elements	Total cost of road elements £m	% of overall multi-modal strategy cost
Access to Hastings (A21 & A259)	December 2000	East Hastings bypass; West Hastings bypass	131	56
Cambridge to Huntingdon (A14)	August 2001	A14 widening	192	74
South East Manchester (Stockport, Manchester airport link west, Poynton)	September 2001	Alderley Edge bypass; Poynton bypass; Stockport bypass; Manchester Airport Link Road	210	20
West Midlands Area (M5/M6 & M42 between M40 & M6)	October 2001	M42 widening J3A–7; Stourbridge bypass; Wolverhampton bypass; Link roads in Black Country	1050	14
West Midlands to North West (M6)	May 2002	M6 widening J11A–20	971	57
London to South West & South Wales (A303, M4)	May 2002	M4 widening J19–20; M4 improvements J15, 16 & 18; M5 widening J16–18/19–20; M5 improvements J16 & 20 Dualling A30 Temple–Bodmin; dualling sections of A303/A358; Blunsden bypass	432	15
A1 (North of Newcastle)	May 2002	A1 dualling Morpeth–Felton; B1341–B1342 dualling	137	63
North/South movements in the East Midlands (M1 J21–30)	May 2002	M1 widening J21a–23a, J25–29; A617 Pleasley bypass extension; Kegworth bypass; Glapwell bypass	611	33

A453 (M1 to J24 (Nottingham))	July 2002	M1 improvements J24; A453 widening M1–Clifton	62	16
Hull (East/West) Corridor (A63 and A1033 to Port of Hull)	July 2002	Upgrading of Castle Street	44	22
South and West Yorkshire	September 2002	M1 widening J30–42; M18 widening J2–3; M62 widening J25–32; A1 widening Worksop; A628 Cudworth bypass; Sheffield-Rotherham link road	687	100
South Coast corridor	September 2002	M27 widening J3–4, J11–12; A27 Worthing–Lancing Tunnel; A27 Arundel bypass, Selmeston bypass, Wilmington bypass	594	54
London to Ipswich	October 2002	A12–M25 to Chelmsford; A12 Chelmsford to A120 Colchester; A12 Colchester to Ipswich; A120 Colchester to A12 Marks Tey	386	65
Orbit	November 2002	M25 widening J5–7, J16–23, J27–31	850	100
Tyneside area	November 2002	A1 western bypass widening; A19 grade separation of junctions; New Tyne crossing	509	40

between Leicester and Chesterfield. In addition to nonroad-based transport measures including improvements to heavy rail through the reopening of the Clowne and Matlock-Buxton lines and the enhancement of existing mainline services, there is a long list of road building measures, including the widening of 37 miles of the M1 between junctions 21A to 23A and junctions 25 to 29, and new bypasses at Kegworth and Glapwell.[40] Somewhat defensively, but nevertheless understandably given the initial objectives of the MMSs, the consultants point out that the road enhancement measures constitute just 31 per cent (£564 million) of the total costs, and claim that this does not amount to a return to previous road building policies. Instead it ' . . . reflect(s) a standard that is needed to achieve the study objectives and to meet the targets set out in the government's Ten Year Transport Plan'.[41] But given that one of the means being deployed to achieve the target of reduced congestion is that of adding more road space, it is difficult to deny that the suggested strategy contains an element of predict and provide.

Details of which sections of the motorway network will be widened are only starting to become available, but reference to the government's 'motorway traffic stress maps' provides a very strong clue as to their likely whereabouts (Figure 4.1). These maps show those stretches of motorway operating at beyond 100 per cent capacity in 1996 and, based upon a 'mid-range' traffic growth scenario, those projected for 2016.[42] Whereas, with the exception of some 'hotspots', virtually the entire motorway network was operating at below capacity in 1996, by 2016 it is projected that all of the M6 and M25, most of the M1 and most of the M62 will be subject to 'stress'. All of these sections are being examined within the context of MMSs and it thus appears that many – such as the M1 through the East Midlands – will be recommended for widening. As an interim measure, pending the outcome of the MMSs, the government announced a £145 million package of engineering works aimed at easing around 140 or so of the most acute bottlenecks on the principal road network through measures such as improved junction alignments.[43] In December 2002, the widening of the M6 between Birmingham and Manchester, and of the M1 between junctions 21 and 30 in the East Midlands, was confirmed.[44]

Interestingly, the draft MMS for the London area, *Orbit: Transport Solutions Around London*, recommends the widening of the M25 notwithstanding Labour's unconditional abandonment of all such plans in *A New Deal for Trunk Roads in England*.[45] As the *10-Year Plan* gives a clear commitment to 360 miles of motorway widening as well as the construction of up to 30 new trunk roads, it is apposite to question the purpose of MMSs since decisions on at least two of their critical components have, in effect, been made in advance of their being undertaken. The referral to separate implementation agencies of the various components recommended by the

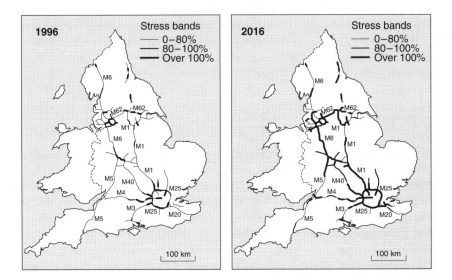

Figure 4.1 English motorway traffic stress in 1996 and projected motorway traffic stress in 2016. The map for 2016 assumes an outcome midway between the least and most stringent implementation of proposed integrated transport policies (see note 42). Source: Department of the Environment, Transport and the Regions (1997) *What Role for Trunk Roads in England? Consultation Paper.* Volumes 1 and 2. The Stationery Office, London.

studies has exacerbated this problem. Whereas road schemes have been referred to the Highways Agency, which has generally welcomed them, proposals for 'heavy' rail upgrades have received a much more guarded response from the Strategic Rail Authority (SRA). In part this is because of funding limitations but also it is because the SRA has its own agenda (Chapters 5 and 10). It is only when the preparation of all MMSs proceeds from a clean sheet of paper, with all proposals being subject to NATA, no elements having been predetermined and each proposal receiving equal treatment from its respective implementation agency, that this new approach introduced by Labour will be capable of producing genuinely innovative and integrated transport strategies.[46]

Finally, although it appears that the multi-modal process could result in a significant expansion of the road building programme, it is open to question as to how many of the schemes will be delivered before 2010.[47] Some of the roads recommended in the MMSs are not scheduled for construction until beyond 2011 in any case, but in anticipation of the likely delays resulting from lengthy planning inquiries for the more urgent schemes, ministers suggested that decisions on major infrastructure proposals should be made by parliament, with public inquiries being to all intents and purposes abandoned. This proposal proved extremely controversial and was dropped. Although the government is intending to introduce procedural changes to speed up the public inquiry system it is most unlikely that these will be capable of reducing delays sufficiently to ensure delivery of all of the schemes by 2010.

Local roads

The expansion of the trunk-road network is to be paralleled by a significant expansion in the local road network. £6.2bn is earmarked in the *10-Year Plan* for spending on local roads 2001/02–2003/04 and altogether around £30bn is provisionally identified for local road spending 2001–10. These planned expenditures, prepared in the light of the first round of bids for transport funding (in the form of Local Transport Plans (LTPs) – see Chapter 3) made to central government by local authorities, provide for the construction of 70 local bypasses and 130 other major local road improvement schemes costing over £5 million. Each year, the government publishes Local Transport Settlements setting out those local transport packages which it has approved for funding. The contents of the successful bids – which, excluding junction improvements, have included 82 road schemes since 1998 – indicate that many local authorities continue to regard road building as the most effective remedy to local congestion problems, not to mention local economic development problems.[48] Moreover, using the emerging Local Transport Strategy (the Scottish equivalent of LTPs) in Aberdeen as a case study, a recent paper has argued that there is enough flexibility within the interpretation of NATA to permit local authorities to include significant road building schemes within 'sustainable' local transport packages'.[49] Environmentalists raised concerns about the scale of local road building approved for England in 1999 and 2000, but gave a guarded welcome to the 2001 settlement which was considered to be more consistent with the ethos of sustainable transport.[50]

Road User Charging and Workplace Parking Levies

Part of the explanation for the rapid growth in car ownership and car usage over the past 20 years almost certainly lies with the reduction in the cost of motoring (Figure 1.4). Reversing this trend through imposing significant increases in fuel duties and vehicle excise duties could go some way to reducing levels of congestion but in practice such mechanisms are seen as being fiscally regressive in that they fail to discriminate between those that can and can not afford them. Labour for two years perpetuated and even increased the 'fuel tax escalator' – introduced by the Conservatives to increase fuel duty annually by five per cent in real terms – before discontinuing the policy after a direct action campaign against the high cost of petrol and diesel in the autumn of 2000 (Chapter 1). It also recognized the potential offered by more discretionary tax measures in the form of road user charging and a workplace parking levy.

Many transport experts have argued that road user charging constitutes a rational, defensible and ultimately inescapable response to the problem of excessive demand being placed upon scarce (but free) road space.[51] In contrast to fuel duties, road user charges can be applied in a targeted and discretionary manner to discourage drivers making journeys on specific stretches of the road network at specific times of the day. Poorer people in rural areas, for whom frequent car use might be unavoidable, would be largely unaffected by the introduction of road user charges, whilst poorer people in urban areas, where congestion is more prevalent, would have greater opportunity to use public transport alternatives. Further, if the revenues generated are used to offset other specific taxes such as fuel duty, then road user charges can be fiscally neutral.[52]

The arguments for a workplace parking levy are more recent and stem in part from concerns expressed in debates during the 1990s about the adverse impact that out-of-town retailing, which is highly dependent upon the provision of free parking, was having on established town centres.[53] A number of interests such as the Council for the Protection of Rural England (CPRE) argued for the introduction of some form of local authority levy on free parking as a 'straightforward way of levelling the playing field'.[54] Although the government rejected the case for applying a tax on free parking at out of town retail complexes, it nevertheless included provision in Part III of the Transport Act (2000) for local authorities to introduce a workplace parking levy.[55] Monies raised through workplace parking levies and road user charges are to be hypothecated for investment in local transportation projects. The *10-Year Plan* assumes that outside of London 8 large towns and cities will introduce road user charging by 2010, and 12 will introduce some form of workplace parking levy.

The application of charging to certain routes within an otherwise free access road network will only be viable in circumstances where a sufficient number of road users takes the view that the monetary benefits of time saved (possibly combined with reduced fuel costs) equal or outweigh the charge. Where these circumstances do not prevail road users are likely to travel on the free network, even if this means 'ratrunning' along unsuitable routes. It is largely for this reason that toll routes in the UK have been restricted to bridges and tunnels across / under river estuaries where there are no obvious alternative routes, although even in the case of the Severn

Bridge there have been claims that toll increases have caused drivers to undertake lengthy detours.[56] Policy makers will monitor closely the level of usage of the 27-mile BNRR after it opens in 2003 as it should provide a useful indication as to how willing road users will be to pay a fee in order to avoid other heavily congested routes.

Until recently, the rudimentary level of technology has meant that road charges have had to be collected manually at toll booths which invariably disrupt traffic flow causing delay and necessitate extra land-take. But with the introduction of cameras capable of scanning registration plates, the need for tollbooths has been dispensed with. In the medium term – certainly, according to the Commission for Integrated Transport (CfIT), within the next 10 years – it will be possible to replace road side cameras with satellite technology (global positioning systems) permitting the introduction of highly sophisticated and discretionary charging regimes at different geographical scales and based on a multitude of factors such as distance travelled, size of engine, level of congestion, time of day and so on.

In the wake of the Transport Act (2000), around 35 local authorities showed an interest in introducing local 'cordon' charging and work-based parking levies, suggesting that the prospects for realising the number envisaged in the *10-Year Plan* were good.[57] But in its report on the *Plan*, the House of Commons Environment, Transport and Regional Affairs Committee concluded that only three cities in England (London, Bristol and Durham) were still pursuing local road user charging and only one (Nottingham) was actively pursuing the introduction of a parking levy.[58] Such a low take-up would appear to jeopardize achievement of the *Plan*'s congestion reduction targets but the government, which has stood accused of failing to take a sufficiently strong lead in promoting road charging for fear of alienating electoral support, rejected this, arguing that the schemes were never of more than marginal importance. This is in stark contrast to the view of CfIT, the government's own expert advisory committee on sustainable transport, which argues that 20–25 per cent of the congestion target is at risk if none of the eight local schemes is implemented.[59]

Quite how widely local road user charging will be adopted will, in very large part, depend upon how successful the schemes now introduced transpire to be.[60] A small scheme was introduced in Durham in 2002, with drivers paying £2 to leave a defined area of the city's historic core. In London, a charge of £5 is made to drivers entering the inner area from King's Cross in the north to Elephant and Castle in the south, and from Vauxhall Bridge in the west to Tower Bridge in the east (Figure 4.2). The Mayor of London, Ken Livingstone, has suggested that the charge could reduce city centre traffic levels by as much as 15-20 per cent, which, if not accompanied by corresponding increases beyond the charge zone, might be sufficient to convince sceptical councillors in other cities to follow suit.

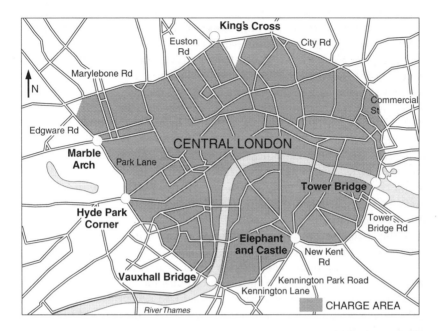

Figure 4.2   The central London road user charging scheme. Source: After Transport for London (2002) Congestion charging . . . where & when does it operate. www.cclondon.com/WebCenterBrandedTR4/ Images/Cntl-Lndn-Congestion-Map-A5.gif (accessed 27 November).

Although Labour has given only lukewarm support to road user charging so far there are signs that, because of anticipated difficulties in complying with new air quality standards introduced in 2002, it has recognized their eventual inevitability.[61] The government's best practice guidance on the preparation of MMSs requires consideration to be given to the introduction of charges for all vehicles travelling on trunk roads / motorways beyond 2010. Each of the MMSs should thus incorporate an analysis of a range of road user charging scenarios, covering different toll levels, different combinations of urban charging and trunk-road charging and so on, in order to assess projected levels of trip demand suppression, modal transfer and trip diversion. The preliminary results of the studies undertaken to date provide something of a mixed picture as to the anticipated impacts of charging.[62] According to the authors of the *London to South West & South Wales Multi-Modal Study*, for example, the main impact of applying tolls of five to 16 pence per mile on the M3/M4 would be 'to cause diversion of highway trips away from the trunk-road network onto 'less suitable' county roads or onto other trunk roads'.[63] On this basis the consultants concluded that the adverse effects of diversion caused by tolling outweighed any benefits that might be achieved. Of course, such a problem would not arise if charges were to be applied on an area wide basis as being proposed in both the *Orbit* and the *South and West Yorkshire* MMSs.

In sharp contrast, the adoption of entry charges of £2.50–£10.00 per day along the M6 between Birmingham and Manchester (the *West Midlands to North West Conurbation Multi-Modal Study*) would, it is estimated, result in projected increases in the number of passengers using public transport alternatives of between 45,500 and 82,800 each day. Similarly, the adoption of road user charging in the *West Midlands Area Multi-Modal Study* is thought to provide 'greater flexibility and mode shift potential than other possible economic interventions . . . [making it] . . . a key part of the . . . strategy . . . to bring the perceived cost of travel by car nearer to public transport travel costs'.[64] The *Orbit* study also recommends the adoption of a charge for all vehicle use within the south east of England once the requisite technology becomes available. But as the authors of the *South Coast Corridor Multi-Modal Study* stated, the key question is, 'at what point do tolls need to be set to achieve levels of modal transfer and will this be politically acceptable to South Coast communities?'[65] It is only if and when these people of 'middle England' come to accept the inevitability of tolls that the government might accept their potential role and legitimacy in transport planning.

Labour's Roads and Traffic Policies in Perspective

When compared against its original commitments to make the construction of new roads an option of the last resort and to reduce the number of car journeys, Labour's record on roads and traffic, which must lie at the very heart of its wider transport policy, has been extremely disappointing. Since 1997, when John Prescott gave his now infamous pledge on traffic reduction (quoted at the beginning of this chapter), the *total* number of car journeys has increased every year although the *rate* of growth, which he subsequently – and dubiously – claimed to be referring to, has slowed down.[66] In this respect Prescott claims that Labour's transport policy has already begun to succeed. But many critics, including the former chair of the House of Commons Transport Select Committee, Labour MP Andrew Bennett, have been scornful, accusing the government of lacking courage in taking strong action against car usage.[67]

These critics are right to raise concerns. For a start, it is difficult to square Labour's pledge to make road building a means of the last resort with the declarations of intent set out in the *10-Year Plan*. When the *lengths* of the targeted programme (323 miles), the motorway widening programme (360 miles) and the multi-modal programme (up to 180 miles) are aggregated it is apparent that the scale of Labour's trunk-road proposals for England up to 2010 – around 85 miles per year – is little different from that achieved under the Conservatives (95 miles per year). Similarly, ministers in Northern Ireland, Scotland and Wales have also launched major trunk-road construction programmes. The Scottish Executive, for example, has recently confirmed that around 60 per cent of its transport budget in the three financial years to 2000/04 will be spent on trunk roads and motorways (Chapter 2).

Of course, the eventual construction out-turn will be critically dependent upon the outcome of the MMSs and the ability of the government to secure sufficient tax revenues in the years ahead to fund such an ambitious programme. With respect to the first issue, those studies that have been published place a significant emphasis upon road building and, to say the very least, it is clear that the integrity of the evaluation process has been severely compromised – and tainted – by pre-existing commitments. In addition to the 360 miles of motorway widening, many of the 44 road schemes being evaluated as part of the MMSs are likely to be recommended and approved for eventual construction. The government might argue that such solutions are practical and perhaps unavoidable – indeed, there are cases where the targeted building of even major new roads can bring real economic, social and (local) environmental benefits – but there is a suspicion that other, more sustainable options, are not being considered in sufficient detail or are even being ignored altogether.[68] Regarding the second issue, recently issued projections point to an emerging public sector deficit which may result in reductions in planned government expenditures – whether they be in health, education or transport – or a need for alternative (that is, private) funding sources.[69] It is distinctly possible, however, that ministers would opt to cut other planned transport investments over the road building programme if faced with a squeeze on public sector finances, as this would probably be the most politically expedient option. Such a policy decision would perhaps be justified by the government on short term value for money grounds.[70]

It must be acknowledged that more than 50 per cent of the total additional mileage will come about through road widening – rather than through the construction of completely new roads – whereas under the Conservatives the corresponding figure was between 2 and 23 per cent. Nevertheless, even with this proviso, it is still the case that Labour is providing for a very considerable level of road building over the course of the *10-Year Plan*.[71] In this sense Labour's policy is little different from that of the Conservatives, although, of course, Labour is also committed to increasing investment in public transport (Chapters 5, 6 and 7). But just like the Conservatives, it appears that Labour dare not upset road users. The government's rather permissive road building policies have been devised to ensure their continued political support which it sees as being vital to its long term political interests.

And just as new roads have not become an option of the last resort, neither has road user charging, arguably the most potent weapon to ensure that there are 'far fewer journeys [made] by car', been embraced with any real political commitment.[72] No doubt recognising that charging was never going to be popular – underlined by the damage that the fuel protestors inflicted upon Labour's standing in the opinion polls in September 2000 (Chapter 1) – the government has handed over responsibility for devising and implementing schemes to local authorities. Notwithstanding their initial enthusiasm for such schemes, in the face of local voter opposition and fearful of losing out economically to neighbouring towns and cities that might opt against charging, many local authorities have subsequently dropped plans to introduce them. Whilst allowing such decisions to be made by local authorities is consistent with the democratic ethos, there must be at least a suspicion that Labour is frightened of being too closely associated with what could turn out to be a massively unpopular policy. On a related note, ministers concede in the *10-Year Plan* that they expect the real costs of motoring to fall further – by around 20 per cent – over the next decade, whereas they anticipate that public transport costs will remain constant. CfIT has estimated that such a reduction will result in appreciable additional traffic growth. Taking other factors into account, traffic is projected to increase by 17 per cent if motoring costs fall by a fifth, as opposed to 13 per cent if the government adopted a policy to ensure the amount paid by drivers remained constant.[73]

If public sector finances deteriorate as projected, the government, assuming it intends to honour its current road building commitments, might have little option but to bring forward charging.[74] There is some evidence that the Department for Transport has become more willing to consider introducing road charging in the medium-, rather than the long-, term future. The current Secretary of State for Transport, Alistair Darling, has called for a national debate on the issue and has approved a £22 million pilot project in Leeds examining how to charge drivers.[75] Importantly, a distance-based charge for freight hauliers has been announced and is scheduled to start in 2006, although the details of the scheme – including the cost to haulage companies – are still to be decided. In putting any motion for charging the government must explain the true financial costs of motoring (accounting for the externalities associated with motoring) and so debunk the widely believed myth that motorists pay more in taxes than they receive in benefits. Simultaneously it will have to advocate the lessons that can be learned from those towns and cities in Europe that are tackling the issue of unrestricted car usage and realising considerable environmental, economic and social benefits for their citizens (Chapter 1).[76]

Whilst Labour's record is one of some disappointment it is not one that should cause much surprise. It is now clear that Labour underestimated the scale of its task and was mistaken in its belief that traffic growth, which has continued at a remorseless rate for decades, could be reversed in just five years without the introduction of extremely punitive measures and vast improvements to public transport. By the time *A New Deal for Transport* was issued in 1998, there was neither mention of roads being a measure of the last resort nor was there any attempt to impose a challenging and clearly articulated national traffic growth target. Although the Road Traffic Reduction (National Targets) Act (1998) commits the government to setting and publishing a target for road traffic reduction in England (and separately in Scotland and Wales), it absolves the Secretary of State of this responsibility should he/she deem an alternative mechanism more appropriate and can justify his/her reasoning. Furthermore, the Act does not require that the targets adopted by the Secretary of State be legally binding, or that he/she should publish 'progress reports' unless he/she deems it appropriate. In reality, therefore, the Act has little real value and allows the government significant future flexibility when considering the desired outcomes of its transport policy.

Labour's objective, it seems, has been to extend transport choice whilst also providing for the more frequent use of the car (Chapter 1). The decision to adopt new measures capable of reducing road building and managing traffic demand – NATA, MMSs and charging regimes – was a clear step forward, but to compromise their delivery in practice and revert to predict and provide style road construction programmes can be regarded, in terms of Labour's initial transport objectives, as two steps back. Irrespective of the ability to fund new roads, for reasons of congestion and pollution continued increases in levels of car usage can not be sustained in perpetuity. Labour will no more be able to build its way out of traffic jams than the Conservatives. Whether public opinion will sway decisively in favour of curtailing road building and the introduction of road user charging and workplace parking levies will, however, depend in large part upon just how tolerant (or inured) people become of congestion. Until such time as this tolerance is snapped, this government will probably seek as far as possible to retain the support of the car owner and user.

NOTES

1 The then Secretary of State for Transport, John Prescott, speaking in 1997, quoted in Friends of the Earth (2000) *Paved with good intentions? Government transport plans*. Press Release, 20 July.

2   Secretary of State for Transport, Alistair Darling, quoted in *Daily Mirror* (2002)   http://www.childrens-express.org/dynamic/public/darlings the driving_27070.htm (accessed 11 December).

3   Department of Environment, Transport and the Regions (1998) *A new deal for transport: better for everyone*. Cmnd 3950, The Stationery Office, London.

4   Department of Environment, Transport and the Regions (1998) *A new deal for transport*, 3.

5   Department of Transport (1989) *Roads for prosperity*. Cmnd 693, HMSO, London.

6   There is, of course, an irony here in that these 'middle-Englanders' are the very people who drive their cars the most (see Chapter 1).

7   Royal Commission on Environmental Pollution (1994) *Eighteenth report. Transport and the environment*. Cmnd 2674, HMSO, London; Standing Advisory Committee on Trunk Road Assessment (1994) *Trunk roads and the generation of traffic*. HMSO, London.

8   Labour Party (1996) *Consensus for change: Labour's transport strategy for the 21st Century*. The Labour Party, London.

9   Department of the Environment, Transport and the Regions (2000) *Transport 2010: the 10-year plan*. DETR, London.

10   Scottish Executive (2002) *Scotland's transport – delivering improvements*. Scottish Executive, Edinburgh; National Assembly for Wales (2001) *The transport framework for Wales*. Transport Division, Cardiff; Department for Regional Development (2002) *Regional transportation strategy for Northern Ireland 2002–2012 / Regional transport programme 2002–2003*. DRD, Belfast.

11   Shaw, J and Walton, W (2001) Labour's new trunk-roads policy for England: an emerging *pragmatic multi-modalism. Environment and Planning A*, 33, 1031–56.

12   A trunk road is officially defined as a road for which the Secretary of State, as opposed to a local authority, has responsibility. In practical terms, trunk roads tend to be arterial or other main routes, including motorways, linking strategic destinations.

13   *Hansard* (1989) Volume 153, 18 May, 482–96.

14   Department of Transport (1989) *New roads by new means: bringing in private finance*. Cmnd 698, HMSO, London.

15   National Audit Office (1993) *Progress on the Department of Transport's motorway widening programme*. HMSO, London.

16   Goodwin, P; Hallett, S; Kenny, P and Stokes, G (1991) *Transport: the new realism*. Transport Studies Unit, University of Oxford, Oxford.

17   Department of Transport (1994) *Trunk-roads in England: 1994 Review*. HMSO, London.

18   Department of Transport (1994) *Trunk-roads in England*, 1

19   Walton, W (1996) Policy changes in the government's road building programme: a U-turn or just an application of the brakes? *Town Planning Review*, 67, 437–56.

20   Bryant, B (1996) *Twyford Down: roads, campaigning and environmental law*. E&FN Spon, London.

21   The Primary Route Network consists of those roads the provide the most satisfactory route for through traffic between two or more places of traffic importance. The network is indicated by green and yellow route signs.

22   Walton, W and Shaw, J (2003) Applying the new appraisal approach to transport at the local level in the UK. *Journal of Transport Geography*, 11, 1–12.

23   Bryant, B (1996) *Twyford Down.*

24   COMEAP (1998) *Quantification of the effects of air pollution on health in the United Kingdom.* The Stationery Office, London.

25   Council for the Protection of Rural England (2001) *Government's slow shift on greening transport.* Press Release 53/01, 13 December.

26   Department of Environment, Transport and the Regions (1997) *New roads a last resort – Strang.* Press release 216.

27   Department of the Environment, Transport and the Regions (1997) *What role for trunk-roads in England? Consultation paper volumes 1 & 2.* The Stationery Office, London.

28   Department of the Environment, Transport and the Regions (2001) *Decision on the first multi-modal transport study announced – government announces transport package for Hastings but rejects by-passes.* News Release 2001/0322, 12 July; Department for Transport (2002) *Minister gives commitment to dualling of A66.* News Release 050, 22 August; Department for Transport (2002) *£5.5 billion package of transport improvements.* News Release 2002/0354, 10 December.

29   *Hansard* (1998) Volume 317, 31 July, 653–76.

30   Confederation of British Industry (2002) *From plan to delivery: time to decide. 10-year plan progress report.* Transport Brief, July.

31   Scottish Executive (2002) *Scotland's transport: delivering improvements.* Scottish Executive, Edinburgh; National Assembly for Wales (2002) *Sue Essex announces Trunk Road programme for Wales.* Press Release, 21 March.

32   Walton, W and Farrington, J (2000) The sustainable transport study for Aberdeen: a pioneering attempt at a 'multi-modal study'. *Environment and Planning C: Government and Policy*, 18, 609–27.

33   Department of the Environment (1997) *The United Kingdom national air quality strategy.* Cm 3587, The Stationery Office, London.

34   Department of the Environment (1997) *The United Kingdom national air quality strategy*, 53.

35   Department of the Environment, Transport and the Regions (1997) *What role for trunk-roads in England?*

36   Department of the Environment, Transport and the Regions (2000) *Guidance on the methodology for multi-modal studies.* The Stationery Office, London.

37   Department of the Environment, Transport and the Regions (1999) *Planning policy guidance note 12: development plans.* DETR, London.

38   Department of the Environment, Transport and the Regions (2001) *Decision on the first multi-modal transport study.*

39   Department of Transport, Local Government and the Regions (2001) *Keeble welcomes recommendations of Cambridge to Huntingdon multi-modal study.* News Release 2001/0540, 13 December; Department of Transport, Local

Government and the Regions (2002) *Minister announces strategy to cut congestion in south east Manchester.* News Release DTLR/NW/TRAN004/2001, 21 March.

40 W S Atkins (2002) *North/south movements in the M1 corridor in the east Midlands multi-modal study.* Final Report, April, W S Atkins Consultants Ltd, Manchester.

41 W S Atkins (2002) *North/south movements,* 106.

42 The DETR modelled numerous scenarios for traffic growth depending upon the extent to which it implements the policies contained in *A New Deal for Transport.*

43 BBC (2002) *Plans unveiled to cut road congestion.* http://news.bbc.co.uk/1/hi/uk/2331701.stm (accessed 30 October, 2002).

44 Department for Transport (2002) *£5.5 billion package of transport improvements.*

45 Halliburton (2002) *Orbit: transport solutions around London – the provisional strategy for 2016.* http://www.orbitproject.com/ (accessed 15 August, 2002).

46 House of Commons (2002) Session 2001–2002, HC 558-I, 27 May, The Stationery Office, London.

47 House of Commons (2002) Session 2001–2002, HC 558-I, 27 May: Confederation of British Industry (2002) *From plan to delivery.*

48 Vigar, G; Healey, P; Hull, A and Davoudi, S (2000) *Planning, governance and spatial strategy in Britain: an institutionalist analysis.* Macmillan, London.

49 Walton, W and Shaw, J (2003) Applying the new appraisal approach to transport at the local level in the UK.

50 Council for the Protection of Rural England (1999) *Government bypasses transport with new road schemes.* Press Release, 16 December; Council for the Protection of Rural England (2000) *New roads eclipse good news on transport.* Press Release, 66/00, 14 December; Council for the Protection of Rural England (2001) *op cit.*; Friends of the Earth (2000) *Bulldozer shadow looms over top wildlife sites.* Press Release, 14 December.

51 Maddison, D; Pearce, D; Johansson, O; Calthrop, E; Litman, T and Verhoef, E (1996) *Blueprint 5: the true costs of road transport.* Earthscan Publications, London.

52 Commission for Integrated Transport (2002) *Paying for road use.* CfIT, London.

53 House of Commons (1994) Session 1993–94, HC 359-I, HMSO, London; House of Commons (1997) Session 1996–97, HC 210-I, The Stationery Office, London; House of Commons (1999) Session 1999–2000, HC 120, The Stationery Office, London.

54 House of Commons (1997) Session 1996–97, HC 210-I, xiii.

55 Transport Act (2000) *Public general Acts – Elizabeth II.* Chapter 38. The Stationery Office, London.

56 House of Commons (1994) Session 1993–94, HC 359-I, 94.

57 *Daily Telegraph* (2001) 2 January, 6.

58 Edinburgh is also keen to introduce road user charging and is due to hold a referendum on the matter in 2003.

59 House of Commons (2002) Session 2001–2002, HC 558i, 22 March, 455.

60 Gray, D and Begg, D (2001) *Delivering congestion charging in the UK: what is required for its successful introduction?* The Centre for Transport Policy, Robert Gordon University.

61 *Financial Times* (2002) 6 August, 3; *The Times* (2002) 6 August, 4.
62 Department for Transport (2002) *MMS Infrastructure charging seminar summary papers*. www.dft.gov.uk/itwp/mmsinfra /06.htm (accessed 30 July, 2002).
63 Department for Transport (2002) *Minister gives commitment to dualling of A66*.
64 Department for Transport (2002) *Minister gives commitment to dualling of A66*.
65 Department for Transport (2002) *Minister gives commitment to dualling of A66*.
66 Department of Transport, Local Government and the Regions (2002) *Traffic in Great Britain: 1st quarter 2002*. Statistical bulletin (02). DTLR, London.
67 *Financial Times* (2002) 12 August, 2.
68 Walton, W and Farrington, J (2000) The sustainable transport study for Aberdeen; Standing Advisory Committee on Trunk Road Assessment (1994) *Trunk roads and the generation of traffic*.
69 *The Times* (2002) 25 October, 25.
70 Glaister, S (2001) *UK transport policy 1997–2001*. Address to the British Academy for the Advancement of Science, Glasgow, 4 September; Leach, G (2001) *More roads and road pricing – the way to go?* Institute of Directors, London.
71 *Daily Telegraph* (2001) 4 January, 4.
72 Quoted in Friends of the Earth (2000) *Paved with good intentions?*
73 House of Commons (2002) Session 2001–2002, HC 558i, 437.
74 *Financial Times* (2002) 21 October, 4.
75 *Financial Times* (2002) 23 July, 2.
76 Hall, D (1998) Urban transport, environmental pressures and policy options. In Pinder, D (ed) *The new Europe*. John Wiley, Chichester, 435–54.

5

A Railway Renaissance?

Jon Shaw and John Farrington

Great Britain is the home of the railway. After George Stephenson's *Locomotion* hauled its first train in 1825 and the first passenger service, between Liverpool and Manchester, began five years later, Britain's railway network burgeoned to a peak of around 23,000 miles by the early 1900s.[1] The train, with its ability to shift people and freight at speeds and in quantities hitherto unimaginable, transformed almost every aspect of Victorian life (it even standardized time), but increasing competition from road transport and the resulting change in spatial land-use patterns have massively reduced the impact and influence of the railway. The network has shrunk to under 11,000 miles and train travel now accounts for just 7 per cent of motorized passenger journeys.[2] As part of a wider package of sustainable transport measures, however, the railway still has an important role to play in the government's policy of integrated transport. Provided genuine modal shift takes place – there is little point in encouraging more rail journeys if road traffic continues to grow strongly as this only generates additional externalities (Chapter 1) – increasing the amount of train travel can help cut pollution and congestion, and confer a range of social benefits.[3]

Although there are obviously environmental impacts associated with rail transport (it mainly uses nonrenewable fuel and 'takes' land in the same way as roads), it is, with the exception of the less flexible and slower modes of water and pipeline, the most space and energy efficient way of moving large volumes of people and freight. To illustrate, a road network needs 13 times more space than a suburban rail network to convey the same number of passengers and trains can be around four times more energy efficient per passenger kilometre than cars in an urban environment.[4] Train travel can also bring economic benefits by reducing road congestion. Indeed, the construction of new rail lines can represent better long-term value for money than road building, particularly on busy inter- and intraurban routes where the scope for achieving modal transfer, and thus for reducing the externalities associated with increasing traffic levels, is greatest.[5] Finally, in social terms, rail transport can help sustain local communities by providing an alternative to private cars for those who cannot access them, and thereby contribute to alleviating exclusion. This is particularly the case beyond the south east of England where more than half of rail users are not in the top two social groups. Rail is, too, the safest form of motorized transport in Great Britain – between 1992 and 2001, there were around eight times as many fatalities per billion passenger kilometres on the roads compared with the railways – and impacts much less on community life and human health than congested roads.[6]

Yet persuading the UK's motorists to help realize such environmental, economic and social benefits by using their vehicles less often would always be difficult in Europe's most car-dominated economy.[7] In 1998 John Prescott, the Deputy Prime Minister and then Secretary of State for Transport, was of the view that the success of his integrated transport policy, outlined in the White Paper *A New Deal for Transport: Better for Everyone*, would depend in part upon the balance of 'sticks' (to penalize excessive car use) and 'carrots' (to make other modes more attractive) it contained (Chapter 1).[8] Introducing 'sticks' such as road user charging might never be politically straightforward, but it would be far less problematic if tangible improvements to rail and bus networks had been, and were continuing to be, made. *A New Deal for Transport* accordingly offered 'the potential for a railway renaissance', and Prescott's then second-in-command, John Reid, suggested that 'if we don't have a decent train system, we can't handle the other problems in creating an integrated transport policy'.[9]

This chapter examines the government's record on rail policy in the context of *A New Deal for Transport* and assesses the likelihood of a railway renaissance – that is, achieving growth targets specified by ministers – occurring by 2010. Discussion focuses on developments only in Great Britain, since Northern Ireland's physically and administratively separate railway accounts for a tiny proportion of rail journeys within the UK as a whole. The limited geographical extent of the Northern Irish network – which will not change in the foreseeable future – and the troubled local political situation with its associated security implications rather constrain the potential for modal shift to rail in much of the Province in any case.[10] The chapter begins by reviewing the policy of rail privatization pursued by the previous Conservative administration because this has heavily influenced Labour's approach to the railway and is of continued relevance with regard to the government's aim of increasing patronage. There is then an assessment of the rail industry's performance under Labour, before a consideration of the extent to which ministers are likely to impel travellers and freight shippers to use the railway more often. Our thoughts on the trajec-

tory of Labour's railway stewardship are presented along with some concluding remarks in the final section of the chapter.

A New Opportunity for the Railway?

The Conservative governments of Margaret Thatcher and John Major
pursued a vigorous policy of industrial privatization which, as other chapters in the book discuss, was extended throughout the transport sector. The
White Paper on rail privatization, *New Opportunities for the Railways*, was
published in 1992. Although it recognized that the efforts of BR and its staff
had contributed to 'significant improvements in recent years', and that 'the
productivity of the BR workforce is among the highest of any European
railway', the White Paper went on to say:

> regular users know that the performance of the railways is not good enough.
> Too frequently, and on too many lines, the quality of service fails to meet the
> travelling public's expectations. BR's staff and management work hard to
> improve services. But they are limited by the structure of the industry in the
> public sector.[11]

New Opportunities for the Railways overlooked the distinct possibility that
sustained underinvestment, rather than 'the structure of the industry in the
public sector', was primarily responsible for the perceived underperform-
ance of the railway. Whilst it pointed out that in 1991 investment was the
highest in real terms since 1962, there was no mention of the fact that
Britain's record on railway infrastructure spending – as with transport
infrastructure more generally – compared poorly with those of other European Union countries (Figure 1.3). It also failed to note that the principal
reason for 'greatly increased' rail investment was the need to prepare for the
commencement of Channel Tunnel rail services rather than the delivery of
strategic, long-term improvements to the domestic network.

After two years of frenetic policy making, a clear structure for the privatized railway industry was finalized in 1994/95.[12] Reflecting the Conservatives' belief in the competitive 'free' market, BR was vertically separated –
that is, the various aspects of its operations (train service provision, infrastructure and engineering and so on) were split apart from each other – in
order to liberalize almost every aspect of the industry. There would be 25
franchised Train Operating Companies (TOCs) to run passenger services,
seven freight operators, three rolling stock leasing companies (ROSCOs)
and a host of maintenance and other support companies. The only
remaining monopoly of any size would be the infrastructure owner, Railtrack, which would sell track access to competing train companies and buy

in the upkeep and development of its network from competing maintenance and track renewal companies. All BR's successor companies would
interact on a commercial basis and relationships between them would be
governed by contract. The Office of Passenger Rail Franchising (OPRAF)
would transfer the TOCs to the private sector, monitor their performance
and disburse subsidy where required. Subsidy was paid only to TOCs –
who would combine this with ticket revenues to pay their bills (such as
track access charges) at market rates and, it was hoped, make a profit – at a
level determined by the franchise bidding competition. The Office of the
Rail Regulator (ORR) would, among other things, oversee commercial
interactions between train operators and Railtrack, and promote competition where possible.

   In the context of this book, it is important to note that *New Opportunities
for the Railways* was conceived neither in the context, nor to achieve the
aims, of a sustainable transport strategy. The Conservatives' tendency to
prioritize road investment throughout the 1980s and early 1990s meant
that the potential for other modes to play a major role in the alleviation of
Britain's transport problems was consistently overlooked (Chapters 1 and
4). *New Opportunities for the Railways* acknowledged that rail transport,
'often cause[s] less environmental damage than road transport', but the
issue of exploiting the railway's environmental benefits featured last in a list
of the White Paper's 'essential priorities' and few words were devoted to
explaining how this might be achieved. Nothing was said about a role for
rail as part of a wider, economic development strategy and social inclusion
was mentioned only in passing: whilst the government 'fully recognise[d]
the social and other benefits of providing regional and commuter services',
and was 'committed to providing continuing subsidy to support them', it
failed to express what it recognized the benefits to be. The tone of the
document, with its emphasis on greater efficiency and value for money,
gave the impression that ministers would not look favourably upon calls for
increasing subsidy to fund the expansion of loss-making but socially necessary services.

   The Conservatives' attitude towards the future of the rail industry also
limited the potential for privatization to accommodate significant growth
and, therefore, modal shift. The sale of BR provided a real opportunity for
the government to influence travel habits. Through franchising, for example,
ministers could have demanded ambitious output targets from the TOCs
and, in theory at least, secured them at a reasonable price from the private
sector. But in addition to factors such as the sheer complexity of the industry,
new contractual arrangements imposed at the time of privatization made it
extremely difficult for train companies and their contractors to accommodate large-scale growth. Access charges paid by TOCs to Railtrack were, for
the first five years of privatization, largely fixed and allowed operators to run

up to 8 per cent more trains than advertised in the timetables they inherited from BR for virtually no additional cost. As things turned out, Railtrack quickly found itself in the position of having to allow many more trains on its network – and accommodate the increased maintenance costs and TOCs' demands for infrastructure enhancement to relieve newly created bottle-necks – without receiving any financial incentive to invest.

The Railway under Labour

In opposition, Labour was against the privatization of BR and made clear in 1995 that it favoured a 'publicly owned and publicly accountable', rail-way.[13] A report published in the autumn of the previous year had referred to the sale as 'fatally flawed', and warned potential buyers that a Labour administration would not be bound by undertakings given by the Conservatives. The then shadow Transport Secretary, Frank Dobson, argued that, 'the whole thing is absolutely hopeless . . . Rail privatisation remains a Tory project which is unsustainable, unworkable and running out of steam'.[14] Although these sentiments appear unambiguous, suspicion grew that Labour, seeking to promote itself as a business-friendly party, might renege on its commitment to renationalize the railway once in power. Indeed, as the 1997 general election approached and it seemed likely that all or most of BR's successor companies would be sold before voting started, Labour confirmed its *volte face*. 'Publicly owned' would, in fact, mean 'privately owned':

> Some people have argued that Labour could have halted the privatisation simply by seeking to repurchase parts of the rail network that have been sold off so cheaply. This is untrue. In the case of Railtrack, a promise to repurchase at anything less than the market price would be illegal under European law. A promise to repurchase at market price would . . . require considerable public resources . . . [and] there would be insufficient resources left for investment.[15]

One writer has argued that Labour, owing to its runaway lead in pre-election opinion polls, could easily have prevented the sale of most of BR.[16] The Party's 1997 election manifesto avoided such a topic, stating simply that 'our task will be to improve the situation as we find it, not as we wish it to be'.[17]

Some 14 months after the general election – rather too long for some commentators – *A New Deal for Transport* was published. Transport Minister John Reid made clear that if the government was to be successful at persuading people to use their cars less often, travellers should have a choice of 'real, safe alternatives to the car', and continued to suggest that '[the] train is the central element in solving all our transport problems'.[18] Reid's assertion was (presumably) not meant to be interpreted literally – a successful integrated or sustainable transport policy will, of course, contain many components working in a complementary fashion – but his rhetoric at least indicated the importance attached by the government to the future development of the railway.

The main railway initiative announced by *A New Deal for Transport* was the creation of a Strategic Rail Authority (SRA) which, subject to instructions and guidance of ministers in accordance with the new integrated transport policy, would 'provide a focus for strategic planning of the passenger and freight railway with appropriate powers to influence the behaviour of key industry players'. The creation of the SRA, which would replace OPRAF and assume some of the Rail Regulator's duties, was generally supported both within and outside of the industry because the lack of a figurehead body under the structure devised by the Conservatives had left the railway open to legitimate charges of 'drifting'.

Unfortunately for those in favour of the SRA's establishment, Labour did not bring the Authority formally into being until 1 February 2001 – it operated in 'shadow' form until that point – on account of delays in preparing the necessary legislation (clauses relating to the formation of the SRA were inserted into the complex Transport Act (2000)).[19] For much of the period 1997–2001, the Secretary of State for Transport found himself in an awkward position with regard to the railway. It was becoming evident that the Prime Minister's office did not share Prescott's enthusiasm for some of the more radical 'sticks' of his integrated transport policy as it did not want the government to be perceived as anti-car (Chapter 1). Prescott was limited in his ability to implement the new transport policy without the Prime Minister's support – in the absence of an Act of Parliament none of the major initiatives which might have promoted modal shift could be introduced – and found working with the privatized railway he inherited problematic.

Whilst repeating his mantra that people should use their cars less and the train more, Prescott did little to encourage this by frequently vilifying train operators, famously referring to them as a 'national disgrace'.[20] He was also facing embarrassing charges from pressure groups that Labour was spending less on the railway than the Conservatives had, a function of the tapering subsidy payments negotiated by OPRAF in advance of the general election. National Rail Summits were organized and, although the need for the industry to work together in pursuit of a better railway was agreed, they were, like Labour's rail policy in general during this period, responsible for few tangible improvements. One analyst has suggested that 'Prescott's attacks on the rail companies were widely seen within the industry as the railings of a powerless man who everyone knew was hamstrung by his leader's fear of alienating business'.[21]

Notwithstanding Prescott's disquiet, the performance of the privatized railway under the first years of Labour's 1997 administration, whilst mixed, was by no means catastrophic. Indeed, in the context of sustainable transport it was rather positive since passenger and freight volumes increased in absolute and relative terms during the mid-to-late 1990s. The number of passenger kilometres travelled rose to a level higher than at any time since the 1950s (Figure 5.1). The amount of freight moved was returning to 1991 levels (the early 1990s were a particularly bad period for BR's freight businesses) and the number of new flows prompted some commentators to suggest that rail freight was heading for something of a revival (Figure 5.2).[22] Although road traffic levels continued to increase, particularly on trunk and motorway routes, the rate of growth may have been less than if the railway had not attracted such a high amount of additional patronage. Some *relative* environmental and economic benefits, however small, may have occurred as a result. If the claim stands up that rail services can help sustain local communities, then the completion of new routes (such as the Robin Hood Line between Nottingham and Worksop) and a series of station reopenings will also have had a positive impact in social inclusion terms.

But just as the industry's changing fortunes were little to do with government policy initiatives, they were only partly associated with privatization. Many of the line and station reopenings since 1997 had been initiated under BR and were completed by Railtrack only with the aid of significant public subsidy. With regard to the train operators, it is true that the freight sector achieved impressive growth rates despite the reduction of 'rail-friendly' heavy industry and manufacturing flows. The initiatives of some

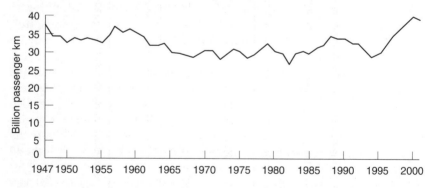

Figure 5.1   Passenger kilometres in Great Britain, 1947 to 2001. Source: Strategic Rail Authority (2002) *National Rail Trends 2001/02, 4.* SRA, London.

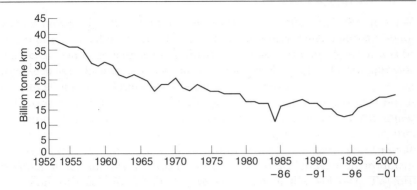

Figure 5.2 Freight tonne kilometres moved in Great Britain, 1952 to 2001/02. Source: Strategic Rail Authority (2002) *National Rail Trends 2001/02, 4.* SRA, London.

TOCs such as innovative marketing campaigns, new services, train refurbishment and the ordering of new rolling stock will also have accounted for additional passenger business (Table 5.1). Significantly, however, the British economy had begun to recover from recession by 1994. The fortunes of the rail industry, like those of most industries, are inextricably bound up with the overall performance of the economy. During recessions, the railway suffers a 'double whammy' effect as fewer jobs in the major conurbations lead to a decline in the number of rail commuters, whilst at the same time demand for leisure travel falls as people change their patterns of expenditure. In the 1990s, passenger numbers had started to rise at least a year before the first TOC franchise was let and continued to increase during the period when most TOCs were either still in public ownership or had not been privately owned for long enough for significant changes in service to have taken place. Although patronage increased more swiftly and consistently after 1994 than in earlier, preprivatization 'booms', it is unlikely that the number of new journeys would have been anywhere near as high in the absence of sustained economic growth.

Indeed, it is difficult to conceive that, had BR been retained and been granted the resources which were devoted to its privatization – selling the company turned out to be extremely expensive – the public sector railway would not have achieved similar levels of growth over the same time period (and made considerable efficiency gains in the process).[23] It is also likely that ministers would better have been able to 'influence the behaviour of key industry players', since the railway would have been under their direct control and primary legislation of the kind needed to establish the SRA would not have been required. Equally, Labour's failure to seize the sustainable transport policy initiative quickly after being elected clearly did little to assist the development of the railway. In this sense, both privatiza-

Table 5.1 Passenger rolling stock ordered/on order since privatization. DMU = Diesel multiple unit. EMU = Electic multiple unit. DEMU = Diesel / electric multiple unit

Company	Type	Order	Value (£m)	Into service	Builder
Completed orders					
Arriva Trains Northern	EMU	8	5	2002	Siemens/CAF
c2c	EMU	184	136	2000–2	Bombardier
c2c	EMU	112	82	2001/02	Bombardier
First North Western	DMU	70	63	2000–2	Alstom
Gatwick Express	EMU	64	45	1999–2	Alstom
ScotRail	DMU	6	7	2001/02	Bombardier
Virgin CrossCountry	DEMU	136	143	2001/02	Bombardier
Total		580	481		
New stock on order					
Anglia Railways	DMU	8	9	2002	Bombardier
Arriva Trains Northern	EMU	8	5	2002	Siemens
Chiltern Railways	DMU	7	8	2002	Bombardier
Connex South Eastern	EMU	210	163	2001/02	Bombardier
First Great Eastern	EMU	84	70	2002	Siemens
First Great Western	DMU	70	63	2001/02	Alstom
Midland Mainline	DMU	127	160	2004	Bombardier
ScotRail	EMU	120	100	2001–3	Alstom
South Central	EMU	240	200	2002/03	Bombardier
South Central	EMU	460	375	2003/04	Bombardier
South West Trains	EMU	120	90	2000–2	Alstom
South West Trains	EMU	785	650	2002–4	Siemens
Virgin CrossCountry	DEMU	216	247	2002/03	Bombardier
Virgin West Coast	EMU	477	592	2002	Alstom
Total		2,932	2,732		

Source: Strategic Rail Authority (2002) *Moving Forward: Leadership in Partnership, Annual Report 2001–02.* SRA, London.

tion and Labour's early stewardship of the railway probably only perpetuated a trend already beginning to establish itself.

A Crisis on the Railway

The launch in July 2000 of *Transport 2010: The 10-Year Plan* announced Labour's transport expenditure plans for the coming decade (Chapter 1).[24] Separate papers advanced the Scottish Executive and Welsh Assembly Government's strategies for their own transport jurisdictions (Chapter 2), although some of the railway funding for Scotland and virtually all of that for Wales was included in *Transport 2010*.[25] The document proposed a considerable increase in funding for the railway. Between 2001 and 2010, it was envisaged that, in cash terms, £26 billion would be spent by the government to lever in a further £34 billion from the private sector.[26] Such investment is rightly expected to produce results and ministers have set targets not only for patronage growth but also modal shift. The SRA was charged with securing an increase in rail passenger kilometres of 50 per cent (80 per cent for inter-urban routes) – the equivalent figure for bus ridership was 10 per cent (Chapter 7) – and in freight tonne kilometres of 80 per cent. It was anticipated that with growth of this order 'people and freight [would shift] off the roads and onto rail', and thereby reduce predicted road traffic congestion (that is, the amount expected by 2010) by at least 3 per cent.[27]

The ability of the railway industry to accommodate new business is obviously associated with the capacity of its network. The need for capacity enhancement was highlighted by the *10-Year Plan*, which estimated that existing constraints would limit passenger growth to no more than 23 per cent above 2000 levels (Figure 5.3).[28] Rail capacity can be increased by using the existing network more intensively (network intensification) and/or extending its geographical scope (network expansion). The first can be achieved through a combination of civil engineering measures designed to remove bottlenecks and accommodate faster and longer trains, such as new signalling, '4-tracking', track renewal and new/extended platforms.[29] Reviewing existing timetables can also increase capacity in some areas by, say, reducing the number of lightly patronized stopping services to make way for additional, heavily loaded fast inter-city services.[30] The second can be realized by reopening disused lines and, in the longer term, by constructing new rail routes.[31] Given that Britain has retained a *relatively* extensive rail network, and also that in some circumstances bus or light rail solutions may often achieve more productive transport outcomes and provide better value for money than building new railway lines, arguably the

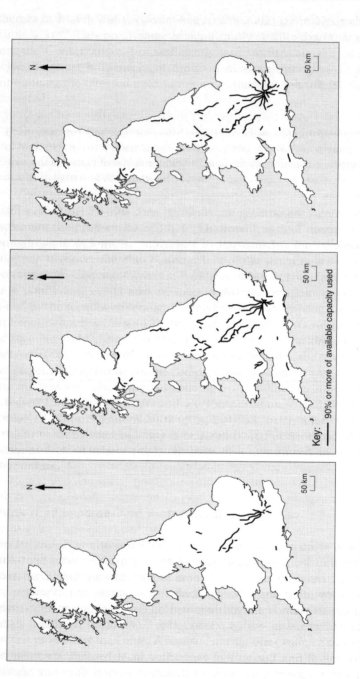

Figure 5.3 Rail lines in Great Britain currently exceeding and forecast to exceed 90 per cent capacity utilisation in 2000, 2006 and 2011. The maps serve as a high level guide only, as actual utilization depends on the exact timetable in place and on other factors. Source: Railtrack (2000) *2000 Network Management Statement for Great Britain*. Railtrack, London.

Key: —— 90% or more of available capacity used

greater scope for capacity enhancement lies in network intensification rather than expansion.[32]

Although by the summer of 2000 the SRA was preparing to formally commence its duties and a greatly enhanced funding package had been put in place, events within the railway industry would conspire to undermine the credibility of the government's new strategy before its scheduled implementation. In a sense, the industry was becoming the victim of its own success. The number of daily passenger services increased to accommodate growth by eight per cent from 17,096 in winter 1995 to 18,470 in winter 1999. Railtrack estimated that, under the prevailing track access agreements, every one per cent increase in the number of rail services was resulting in a 2.5 per cent increase in congestion-related delays as the network became increasingly overcrowded. A review of Railtrack's financing by the Rail Regulator introduced greater flexibility into the access charging regime and resulted in a 35 per cent increase in the company's income (with an additional 5 per cent real terms increase each year until 2005), but the shadow SRA had formed the view that even planned network intensification schemes would be difficult to undertake whilst Railtrack was subject to restrictions on the revenue it could raise from train operators. As a result, the Authority sought to press ahead with TOC refranchising at the earliest opportunity.

Experience from the first franchising round had shown that TOCs with longer (15-year) contracts were more likely to invest in their businesses than those with shorter (seven year) agreements because of the additional time available for amortization (although this was almost inevitable as large-scale investment was usually a precondition of longer franchises being awarded). The shadow SRA invited incumbents with seven year contracts to surrender their franchises and enter into new bidding rounds with the prospect of securing 20-year deals – perhaps with, at least initially, a higher amount of subsidy – contingent upon the promise of massively increased investment programmes. As part of these programmes, it was expected that bidders would explore the possibility for partnership investment in infrastructure projects through so-called Special Purpose Vehicles (SPVs) with Railtrack and other interested companies. Any additional subsidy provided by the public sector could thus be used not only to help fund new rolling stock and other service quality improvements, but also to lever in further private sector investment in the railway infrastructure, over and above that which Railtrack alone could afford.

Refranchising began positively but initial optimism generated by the shadow SRA's policy initiatives would prove misplaced. Since the first of three Heads of Terms agreements was signed in August 2000, the Authority has relet only one TOC, Chiltern Railways. In part the delay has been due to SPV plans being hampered by unforeseen complexities, but progress was also stalled when the prospective fortunes of the rail industry – and thus potential investors' confidence – nose-dived following a major crash at Hatfield, near London, in autumn 2000. The Hatfield crash was caused by gauge corner cracking, a form of metal fatigue, along a stretch of line which Railtrack and its maintenance contractors knew to be faulty and should have subjected to speed restrictions. The fragmented nature of the privatized rail industry has been identified as a contributing factor to the accident because it hindered effective communication between the owner of the infrastructure and those paid to ensure its upkeep.[33] Railtrack's management team imposed emergency speed limits across the whole of the national network and, in so doing, restricted the ability of train operators to run to their advertised timetables (Figure 5.4). Freight movements were badly disrupted, passenger numbers fell by up to 40 per cent almost overnight (they have now more or less recovered) and many TOCs and freight operators were financially imperilled by collapsing revenues.

Railtrack was obliged to compensate train operators for their losses and embark upon a National Recovery Plan to regain the basic standards of engineering which had slipped under its stewardship.[34] The cost of these commitments seriously undermined Railtrack's finances, devastated its share price and compromised its capacity to raise capital (the company's ability to borrow money at reasonable rates was important because the *10-Year Plan* required it to contribute towards the envisaged £34 billion of private sector investment). At the same time, a report on train protection following two earlier fatal accidents – in which, again, privatization was

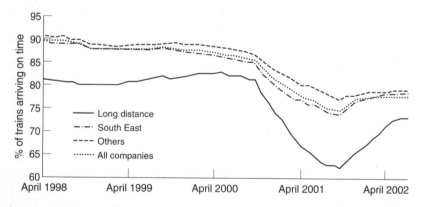

Figure 5.4   Public performance measure (percentage of trains arriving 'on time' – that is, within 10 minutes of timetabled arrival time for long distance operators and five minutes of timetabled arrival time for London & South East and Regional operators), 1998/99 to 2001/02. Source: Strategic Rail Authority (2002) *National Rail Trends 2001/02, 4.* SRA, London.

implicated – recommended expensive safety improvements to the net-work.[35] These factors forced Railtrack to abandon several major infrastructure enhancement schemes which would have provided some of the additional capacity needed to achieve the government's growth and modal shift targets. In point of fact, questions had already been raised over whether the company would ever have been able to carry out its full investment programme as the cost of high profile intensification projects such as the upgrade of the West Coast Main Line (WCML) had already begun to escalate markedly (some would say spiral out of control), from £2.3 billion to over £10 billion. This was due partly to increased transaction costs associated with the track authority industry structure – each industry participant needs to trade profitably – but Railtrack's ineffective project management also contributed. Respected railway journalist Roger Ford has calculated that Railtrack was spending more than twice as much on infrastructure upgrades as BR paid for their equivalents.[36]

Notwithstanding its scaled-down investment commitments and the ORR's revisions to the access charging regime, Railtrack was forced to request direct financial support from the government, in addition to the indirect support it was already receiving via the subsidy paid to the TOCs by the SRA, to stave off insolvency. One and a half billion pounds of funding that was not due to be paid until 2006 was handed over, subject to a string of conditions, including the appointment of a government-approved board member and Railtrack surrendering its responsibility for taking forward future 'grand' schemes. In October 2001, however, the company admitted that even this amount would be insufficient to keep it in business and the new Secretary of State for Transport, Stephen Byers, decided to withhold further assistance. To the delight of many Labour MPs (and others) but the consternation of its shareholders and the City, Railtrack was placed in administration under the terms of the Railways Act (1993). In a separate development, Byers announced the effective post-ponement of much of the SRA's stalled refranchising programme in an attempt to secure 'shorter term improvements that can make a real difference to passengers', through the extension of existing contracts by up to two years.[37]

The decisions to withdraw support for Railtrack and change refranchising policy resulted in considerable uncertainty regarding the direction of the government's post-*Transport 2010* railway strategy. Despite ministers' stated intention to work with the industry structure they had inherited from the Conservatives, newspaper reports began to suggest that officials were now reviewing the potential for fundamental reform. This was hardly surprising. An industry structure closely associated with three significant accidents – although it is important to note that, overall, the privatized railway is statistically no less safe than its nationalized predecessor – and incapable of accommodating at a reasonable cost the increases in passenger and freight traffic which had already taken place, not to mention those which are now required by the *10-Year Plan*, could obviously be judged as hindering the potential for a railway renaissance. Transport academic and former BR manager Lord Bradshaw's 1991 prediction that the track authority model of rail privatization would be expensive, unsuitable and unsafe seems with hindsight to have been all too prescient.[38]

Had the level of postprivatization growth not been so great and had the Hatfield disaster not occurred, the current industry structure might not have been subject to such scrutiny – at least in the short-to-medium term – as it could have coped with more additional traffic and would not have destroyed the confidence of would-be investors. As things stood, however, the fortunes of BR's successor companies would very much depend on how Labour's railway policy developed to address what had quickly become known as an industry crisis. It now appears that major structural change has again been ruled out, although a number of initiatives were unveiled by the government and its regulators with the potential to create a significantly more stable operating environment to encourage growth.

Rethinking Railway Policy

In May 2001 the SRA issued its *Freight Strategy* which recognized the challenge of the government's rail freight targets and set out a generally market-orientated approach to achieving them.[39] The *Strategy* was important because, prior to the SRA's existence, the freight sector was poorly represented within the broader railway industry on account of the relative dominance of passenger operators. The SRA's determination to increase rail freight tonne kilometres – manifested in specific policies, grants and plans for infrastructure improvements to support the private sector in its development of new services – was augmented by the ORR's autumn 2001 announcement that track access charges for rail freight operators would be cut by half as of the following April in an effort to make the sector more competitive with road haulage. The annual subsidy of around £100 million required to make up the shortfall in Railtrack's access charges is provided by the SRA.

In December 2001, slightly in advance of losing its shadow status, the SRA unveiled a new passenger franchising policy. Among a series of measures, it was announced that certain franchises would be consolidated for efficiency reasons, particularly those which shared London termini, as the Authority estimated that up to 10 per cent of potential capacity is currently being wasted by allocating dedicated platforms to individual operators. Particularly notable was the SRA's changed attitude towards the role of private sector innovation in the identification and development of major railway projects. The shadow SRA's first chairman, Sir Alastair Morton, maintained that the Authority should leave franchise bidders to outline their visions for new services and award contracts to those it considered the most competitive. This approach proved unworkable in practice. Not only did it render the SRA unable to plan strategically – this responsibility in effect passed to private companies bidding for individual franchises, rather than the network as a whole – it also made the task of deciding between different franchise proposals extremely difficult. The SRA's dilemma was compounded by the fact that it had failed to produce robust criteria against which it could judge the bids. An industry joke at the time suggested that Morton's organization should be renamed 'R' as it possessed neither Strategy nor Authority.

Morton's successor, Richard Bowker, opted instead for the far less complex approach of identifying a series of core outputs for each franchise which all bidders would have to include in their submissions. Potential franchisees could provide a 'menu' of innovations from which the SRA could choose should it so wish. Bowker's ideas were clearly designed to reclaim a substantial degree of control over the future development of the railway industry, and more of his strategy was forthcoming in the *Strategic Plan*, published in January 2002.[40] The *Plan* announced an additional £4.5 billion of public sector funding over the period until 2010 and, alongside a series of platitudes regarding the need to deliver a safe, reliable, value for money railway 'fit for the 21st century', noted the importance of delivering the 'basics' effectively on a daily basis. This recognition that the industry would have to 'walk before it could run' was particularly apposite given the extent to which serious weaknesses in the structure and functioning of the industry had been exposed. The SRA has subsequently announced that its priorities will be best achieved by letting shorter franchises, almost completely removing from TOCs any innovatory role (the train operators are no longer to think up large scale investment schemes as the Authority has taken this task upon itself), and imposing upon franchise holders much more rigorous quality stipulations covering issues such as punctuality, cleanliness and passenger information. These post-*Plan* developments represent the final nails in the coffin of Morton's vision of an entrepreneur-led railway by turning TOCs' franchise agreements into little more than tightly controlled service delivery contracts.

A further development occurred in March 2002 when Network Rail, the not-for-profit company which with the support of the government has taken over from Railtrack as the network operator, announced a £9 billion funding package. Most of the money, around £6.5 billion, will be used to cover Railtrack's debt and the remainder will be split between working capital (£2 billion) and payments to Railtrack shareholders (£500 million).

Network Rail was assisted in raising its funds by 'a "nod and a wink" financial guarantee from the Treasury', and subsequent announcements have revealed that total government backing could be as high as £21 billion.[41] The promise of Treasury guarantees – which some have interpreted as the effective renationalization of Railtrack – represents a highly significant change in railway policy. Successive governments had been reluctant to underwrite loans as they would count against the Public Sector Borrowing Requirement (PSBR), but the new arrangement will allow Network Rail to raise necessary capital quickly and relatively inexpensively, particularly in an unfavourable investment climate.[42] Interestingly, ministers have been unwilling to countenance a similar means of paying for much-needed investment in London Underground (Chapter 6).

Taken together, these policy initiatives indicate that the government recognized it would have to assume a more active role in the operation, development and financing of the railway than it had foreseen in the *10-Year Plan*. In the wake of the events of autumn 2000, ministers seemed to acknowledge the need to demonstrate an increased level of financial and regulatory responsibility for both the current standard of train services and their future improvement if the *Plan*'s patronage growth and modal shift targets were still to be met. *Rail* editor Nigel Harris has remarked that with renewed government support, the railway had 'a very worthwhile "edge" for the first time in many years'.[43] Yet even assuming that progress in the development of a revised rail strategy returns organizational and institutional stability to the industry, and that the UK economy continues to avoid recession, there are at least five possible reasons why Labour might not be able to meet the *10-Year Plan*'s growth targets by 2010. The first three pertain specifically to internal industry matters, whilst the remaining two are related to developments elsewhere in Labour's transport policy.

Incredible Targets

First, notwithstanding the likely implementation of 'cost-free' measures such as timetable overhaul, the amount of investment required to ensure a sufficient increase in capacity will be significantly in excess of that budgeted by the government. Reasons for this include continued project-cost inflation (Railtrack's successor will struggle to contain costs in the immediate future), Network Rail's assets being in a worse state than previously thought and higher-than-expected subsidy requirements (11 out of the 25 TOCs are making losses after payment of subsidy despite the high growth rates over the last six years, and ailing TOCs will have received around £650 million more from the government than they were originally entitled to by 2003/04).[44] Indeed, Richard Bowker has twice been forced to scale down the SRA's capacity enhancement plans since December 2001. There has also been speculation that the amount of public money actually available to the railways could be less than the headline figures suggest owing to an element of 'double counting' in ministers' presentation of the funding package.[45] Should this prove to be the case, the SRA could face the dilemma of having to pay an already-inflated amount for the same outputs with less money.

Second, the private sector may not be willing to invest anything like the amount projected in the *10-Year Plan*. This is partly because no-one has a complete understanding of the condition of infrastructure assets in the continued absence of a reliable 'asset register'. Investors in the post-Hatfield railway will rightly demand a reasonable guarantee that such an event is far less likely to occur again. Their confidence may have been undermined by another serious train crash in May 2002 at Potters Bar, to the north of London, following a points failure which could have been the result of poor maintenance (at the time of writing, the cause has still to be confirmed). Would-be franchisees, perhaps already concerned by the financial plight of many existing TOCs, will also be concerned about sharply rising labour costs and, following a series of strikes, apparently worsening industrial relations which could further diminish the prospect of profitable operations in the future.[46] Although the government has shown a willingness to provide ailing TOCs with extra subsidy rather than risk the service disruption which may result from a bankruptcy, recent events indicate that such largesse is about to end, with ministers cutting £312 million from the SRA's budget in December 2002.[47]

Third, and aside from funding, is the issue of skills. Rationalization programmes following privatization have resulted in a shortage of technicians and engineers in the rail industry. Drivers, too, are in heavy demand and this is being reflected by sizeable increases in their pay. The SRA has responded by announcing the establishment of a National Rail Academy in its *Strategic Plan*, but whether this will be sufficient to offset the current 'brain drain' is debatable. Answering an MP's question about the implications of a skills shortage for the realization of the proposals in the *10-Year Plan*, David Begg, the Chair of the Commission for Integrated Transport (CfIT), stated:

> [T]he evidence we [CfIT] are gathering now is that it is the physical resource constraint which is greater than the financial resource constraint . . . I do not have an answer to this but a question which interests me is, are we going to spend £180 billion in the *10-Year Plan*? When people say, 'Is it enough?' my first question is, 'Can we make sure we have the physical resources in place to make sure we are going to spend that given amount of money?'[48]

Fourth is the potential impact of airline competition. The *Rail Industry Monitor* suggests that the advent of the 'low-cost' carriers has already affected the patronage levels of intercity operators (passenger numbers remain below their previous peak in 1988/89) and the probable expansion of the UK domestic air market – Labour is likely to increase airport capacity greatly over the coming 30 years (Chapter 9) – could exacerbate the plight of the long-distance train companies.[49] Airline competition could be particularly significant in terms of the *10-Year Plan*'s growth targets for rail since the government expects inter-city operators to make a disproportionately large contribution to achieving them (inter-urban passenger kilometres are projected to increase by 80 per cent). Given also that 'short-haul air journeys have higher [carbon dioxide] emissions, per passenger km, than high-speed rail', Labour's expansionist air transport policy would seem to contradict the (limited) environmental aspirations contained within *A New Deal for Transport* (Chapter 1).[50]

Fifth, there is roads and traffic policy to consider (Chapter 4). The Department for Transport's (DfT's) recent decision to allow 44-tonne lorries onto British roads could affect rail freight loadings, although the halving of track access fees for rail freight operators might offset the impact of larger vehicles. A considerable boon for the road haulage industry could, however, result from the introduction in 2006 of a new road user charge to replace the current system of vehicle excise duty and fuel tax. Ministers also anticipate that the real-terms cost of other motoring will drop by 20 per cent over the next decade – whereas public transport fares are set to remain constant – and Begg is of the opinion that such a reduction will result in a considerable amount of additional car use.[51] With regard to road building, Labour announced in 1997 that, in solving transport problems, 'we see new roads as a last resort rather than a first'.[52] It was, therefore, surprising that the *10-Year Plan* contained proposals for the construction of around 85 miles per year of trunk-road schemes – the 1979–97 Conservative administrations' annual average was about 95 miles – and up to 200 major local road improvement schemes. Finally, the vast majority of local authorities in England are not actively seeking to adopt demand management measures such as road user charging or a workplace parking levy, both of which were authorized by the Transport Act (2000). The number of additional car journeys being made – which could potentially have been made by rail or not at all – in the absence of widespread demand management strategies is expected to be much higher than it would be following their implementation.

All of these factors raise important questions over whether the government's emerging rail strategy will be capable of meeting the relevant targets for patronage growth and modal transfer set out in the *10-Year Plan*. Tellingly, the SRA has revised downwards the passenger growth figure from 50 per cent to between 40 and 50 per cent. Bowker explained that 'the target, rather than become a spot estimate of one number, has become a range', and, whilst insisting that '50 per cent remains an achievable target', he acknowledged that 'we [the SRA] felt that it was appropriate, given perhaps the uncertainty in some economic assumptions, actually to put in a range'.[53] That the bottom end of this range is fully 10 percentage points below the target in the *10-Year Plan* would seem to indicate that Bowker is less confident of achieving 50 per cent growth than he was prepared to admit (if 50 per cent is such an achievable target, a more appropriate range might have been 45–55).

If Bowker does entertain doubts, he is probably right to do so. It is probable that there are insufficient levels of finance and skills to provide significant additional rail capacity, and Labour's wider transport policy does not seem helpful. Allowing the expansion of domestic air services, amending charging regimes for road hauliers, permitting a real-terms reduction in motoring costs, pursuing an ambitious road building programme and failing to deliver planned-for demand management schemes around the country hardly constitute a range of complementary measures to promote modal shift. In short, the SRA faces a stiff challenge if it is to meet the *10-Year Plan*'s patronage growth and modal shift targets. *If* the government remains determined that they shall be achieved (recent announcements imply that enthusiasm is waning), it is likely that ministers will be forced once again to revise their approach to rail policy either directly – by further increasing state expenditure on, and influence over, the industry – or indirectly – by tempering their enthusiasm for road and air transport development – over the next few years.

A Railway Renaissance?

In the light of the 'crisis' in the railway industry and the potential difficulties now faced by Labour as it seeks to move train travel up its sustainable transport agenda, it is ironic that the Conservatives should ever have thought that the development of BR was limited by 'the structure of the industry in the public sector'.[54] The legacy of rail privatization is not particularly good: although some improvements have resulted, such as the amount of new rolling stock entering service, the newly fragmented structure of the private sector railway has contributed to, among other things, a lack of strategy, fears (sometimes unwarranted) over safety, minimal cost-effective infrastructure investment and horrendous project cost inflation. Some – including, no doubt, incumbent ministers – might argue that the current state of the railway is entirely the fault of the previous government, whose radical rail experiment went badly wrong: Labour has been left to pick up the pieces of events beyond its control and, in so doing, recognized just in time that regulatory and organizational change, along with significantly increased funding, will be necessary if there is to be any chance of bringing about a railway 'renaissance'.

Certainly the government is to be congratulated for demonstrating a willingness to involve itself more actively in the development of the railway, but the above would be a rather charitable interpretation of events. In the context of recent and predicted future growth in rail patronage, the Conservatives' complex method of transferring BR to the private sector has proven unfit for purpose. But the fact that Labour did not seek to change matters much sooner is both surprising and disappointing. The Party had opposed rail privatization from the outset (it wanted a 'publicly owned and publicly accountable' railway) and identified weaknesses in the new industry structure (the Conservatives' plans were described as 'fatally flawed'). In government Labour failed to seize the policy initiative to address quickly any of the industry's major problems. Effectively, ministers who saw the train as 'central to solving all our transport problems', were content to rely on an industry structure they had heavily criticized to deliver the level of service improvements necessary to place the railway at the heart of their sustainable transport policy. How can they have developed such faith in their predecessors' policy experiment so quickly?

Critique with the benefit of hindsight is easy and possibly unfair, but in this case it would seem reasonable to suggest that if the newly elected government was so convinced that the Major administration's model of rail privatization lacked merit, then its ministers should have acted quickly to address this (they had, after all, had five years in opposition to come up with a credible alternative). Rail privatization had been the most unpopular industrial divestiture ever undertaken in Britain, so it is likely that swift and decisive action would have been voter friendly.[55] Renationalization might have been politically impossible because of the potential for adverse reaction in the City, and large-scale structural change so soon after the Conservatives had implemented their plans could have been counterproductive. Managers and staff can easily be distracted from their main task of running the trains in a climate of continuous reorganization. In reality, though, neither of these extreme options was inevitable. Academics, commentators and industry personnel have suggested various means of improving the railway by addressing some of the more obvious faults in its structure without radical upheaval. Simplification has been a recurring theme, and initiatives such as reducing the number of franchises and streamlining maintenance procedures which are only now being adopted – six years after Labour came to power – could have been introduced reasonably quickly after the 1997 general election, without the need for primary legislation. Those measures such as the creation of the SRA which did require an Act of Parliament were delayed owing to the lack of urgency in pushing the Transport Bill through the House of Commons.

It is suggested elsewhere in this book that the government chose not to prioritize the rapid development of the railways – and, indeed, nonroad transport more generally – after it came to power because of the Prime Minister and others' desire not to be seen as anti-car. Despite the unpopularity of rail privatization, ministers judged that moves to promote an overtly sustainable transport agenda through, say, a massive increase in public sector rail spending could have backfired if the commissioning of major railway projects was interpreted by the electorate as 'wasting' money which could have paid for roads or other public services. Labour's commitment to stay within Conservative spending limits for its first two years in office ruled out large overall increases in public spending, but even if this pledge had been discarded major new expenditures might have been seen as necessitating direct tax increases, which ministers seem to view as electoral suicide. Labour's mandate was in fact so overwhelming that it could have exploited the national 'mood for change' to push through additional railway – and, indeed, other – spending instead of succumbing to what economist Milton Friedman has termed the 'tyranny of the status quo'.[56] Engendering change and increasing investment in the railways would only have been to follow through policies which Labour had espoused during its time in opposition and which seemed popular enough then.

The government's obsession with private sector funding also sheds some light on the decision not to tinker too overtly with the structure of the railway industry after 1997. It is probable that John Prescott supported (after some initial equivocation) the shadow SRA's pursuit of refranchising because the Treasury saw it as a way of minimising the impact of major railway investment on the PSBR, even after the self-imposed period of restricted spending had passed. Government 'interference', so the argument went, might have deterred potential private investment which was of fundamental importance to the realization of a better railway in the future. It is interesting to note that the government has not shown such a willingness to extend private financing to road building: in the *10-Year Plan*, the public/private split for future rail investment is roughly 50/50, whereas for trunk roads it is about 88/12. A cynic might suggest that ministers wanted to use tried and tested methods of funding for road building to make certain they are completed as quickly as possible – in order to assuage growing discontent among the motorists of 'middle England' (Chapters 1 and 4) – but that experimentation was acceptable for the railways.

Ironically, the political fallout from not having engaged with the railway industry quickly or effectively enough is likely to be greater than if Labour had, in the words of *Rail* columnist Christian Wolmar, 'grasped the nettle', in the first place.[57] As both the road and the rail networks become more congested, and related environmental, economic and social externalities increase, there is still no hope of a major improvement in the quality of train travel in the immediate future. Indeed, enhancements undertaken now will simply come on line five years later than was necessary. What is more, there are direct financial implications. The government has ended up having to pay significantly more than it did in 1997 for effectively the same railway (or arguably a worse railway since the state of the infrastructure has deteriorated). True subsidy – that is, the money paid to Network Rail (the company has assumed Railtrack's new access charge regime and receives specific grants for the WCML and other projects) plus TOCs' subsidy payments – will be £3.84 billion in 2003/04.[58] This is more than double the amount received by BR in its last year of operation (when the economy had not fully recovered from the recession) and the difference between the 1997/98 and 2003/04 subsidy payments could fund the equivalent of around 2,000 new railway vehicles. The cost of future investment is also likely to be much higher than it might have been under a different industry structure.

The future of the railway

Many advocates of sustainable transport, whilst welcoming the tenor of the *10-Year Plan*, would view the rail targets contained within the document as interim – a doubling or trebling of current patronage levels might be sought – and, given the probability that these will be missed, will lobby the government to further increase its railway expenditure. Yet if Labour remains reluctant to pursue policies which might be construed as 'anti-car', a further boost to railway investment is unlikely. Commentators such as Stephen Glaister and David Newbery have argued that, since rail accommodates only seven per cent of journeys, too much is already being spent on the network, and prioritising further road construction would in many cases be far more economically and socially justifiable.[59] Given that ministers expected the £64 billion planned rail expenditure to encourage enough modal shift to abate predicted congestion by only 3 per cent – not, in the final analysis, a huge amount – and that the resulting reduction in carbon dioxide emissions was anticipated to be even less, such a position will no doubt find favour with many.

Disagreement over the most appropriate modal share of future transport expenditure is likely to continue and it would seem sensible for the government to consider in some detail the kind of role it wishes the rail industry to play. What are the railways for? Should the view prevail that 'National Rail' (as the former BR network is now known) forms such an indispensable part of a sustainable transport policy that the scope of the targets in the *10-Year Plan* is insufficient, the thrust of transport policy will need to shift again.

Greatly increasing rail capacity would incur an extremely high (in fact, virtually unimaginable in the context of current British spending levels) cost and would take several decades to realise. An alternative view might involve the limited resources currently available being differently targeted, perhaps on maintaining and renewing, rather than enhancing, the existing network, or on the development of a number of high-density commuter and 'super corridors' at the expense of rural networks to alleviate road congestion hotspots. Such options might cost less and would be quicker to implement but could jeopardize the attainment of a continually improving system of sustainable transport.

Debate over the purpose of the railways would play an important part in determining the long-term size and shape of the network in Great Britain. But it is clear that the government faces numerous obstacles in attempting to meet the targets for rail patronage growth and modal shift set out in the *10-Year Plan*. The privatized rail industry, perhaps unexpectedly, contributed to achieving the aims of a sustainable transport policy by increasing its patronage levels in both absolute and relative terms. Despite ministers' wish to build on this accomplishment by facilitating a 'railway renaissance' as part of Labour's *New Deal for Transport*, there are serious doubts that this can be achieved in the terms set out in the *10-Year Plan*. Indeed, it may well be that rail will be unable to play a reasonable part in ameliorating the adverse effects of high traffic volumes across much of Britain by 2010 and beyond.

NOTES

1   Dyos, H and Aldcroft, D (1971) *British transport: an economic survey from the seventeenth century to the twentieth*. Leicester University Press, Leicester.
2   Department of Transport, Local Government and the Regions (2001) *Transport statistics, transport trends – indices data tables*. www.transtat.dtlr.gov.uk/tables/2001/tt/contents/index/tttables.htm#s2 (accessed 29 May 2002); Railtrack (1996) *Network Management Statement 1996/97: investing in Britain's railway*. Railtrack, London.
3   See Potter, S and Enoch, M (1997) Regulating transport's environmental impacts in a deregulating world. *Transportation Research Part D: Transport and the Environment*, 2, 271–82.
4   Tolley, R and Turton, B (1995) *Transport systems, policy and planning: a geographical approach*. Longman, Harlow.
5   OXERA (2000) *The wider impacts of rail and road investment*. The Railway Society, London.
6   Department for Transport (2002) *Transport statistics Great Britain: 2002 edition*. DfT, London. http://www.transtat.dft.gov.uk/tables/tsgb02/text/tsgb.htm (accessed 25 November, 2002); Steer Davies Gleave (2002) *The case for Rail*.

Transport 2000, Association of Train Operating Companies, English, Welsh and Scottish Railway, Railway Industry Association, The Railway Forum, Rail Passengers' Council, Passenger Transport Executive Group; Glaister, S (2001) *UK Transport Policy 1997–2001*. Address to the British Association for the Advancement of Science, Glasgow, 4 September; Shaw, J, Walton, W and Farrington, J (2003) Assessing the potential for a 'railway renaissance' in Great Britain. *Geoforum*, 34, 141–56.

7   Commission for Integrated Transport (2001) *European best practice in delivering integrated transport: key findings*. http://www.cfit.gov.uk/research/ebp/key/index.htm (accessed 29 October 2002).

8   Department of the Environment, Transport and the Regions (1998) *A new deal for transport: better for everyone*. Cmnd 3950, The Stationery Office, London.

9   Quoted in *Modern Railways* (1999) Direction – the political imperative. 56, 37–9.

10   The Department for Regional Development, Northern Ireland, has produced its own documentation in which it outlines its vision for the Province's railway. See Department for Regional Development (2002) *Regional transportation strategy for Northern Ireland 2002–2012*. DRD, Belfast.

11   Department of Transport (1992) *New Opportunities for the railways. The privatisation of British Rail*. Cmnd 2012. DoT, London, 1.

12   The Conservative Chairman of the House of Commons' Transport Select Committee remarked in some desperation: 'The Secretary of State, Minister of State and Department [of Transport] officials have appeared before this Committee and . . . it seems to me that none of them, quite frankly, have a clue about how all this is going to be worked out'. See House of Commons (1992) Session 1992–1993, HC 246iii, 4 November, 39–65; Department of Transport (1994) *Britain's railway: a new era*. Department of Transport, London; Freeman, R and Shaw, J (2000) (eds) *All change: British railway privatization*. McGraw-Hill, Maidenhead.

13   *The Times* (1995) 5 October.

14   *The Guardian* (1994) 17 October.

15   Labour Party (1996) *Consensus for change: Labour's transport strategy for the 21st century*. Labour Party, London, 20.

16   Wolmar, C (2001) *Broken rails: how privatization wrecked Britain's railways*. Aurum Press, London.

17   Labour Party (1997) *New Labour: because Britain deserves better*. http://www.psr.keele.ac.uk/area/uk/man/lab97.htm (accessed 31 July, 2002).

18   *Modern Railways* (1999) Direction – the political imperative.

19   Transport Act (2000) *Public general Acts – Elizabeth II*. Chapter 38. The Stationery Office, London.

20   Quoted in Wolmar, C (2001) *Broken rails*, 111.

21   Wolmar, C (2001) *Broken rails*, 111.

22   Clarke, J (2000) Selling the freight railway. In Freeman, R and Shaw, J (2000) *All change*; Woodburn, A (2001) The changing nature of rail freight in Great Britain: the start of a renaissance? *Transport Reviews*. 21, 1–13.

23 For a discussion about the costs of rail privatization, see: Wolmar (2001) *Broken rails* and White, P (1998) Rail privatization in Britain. *Transport Reviews*, 15, 109–30.

24 Department of the Environment, Transport and the Regions (2000) *Transport 2010: the 10–year plan.* DETR, London.

25 Subsidy for the ScotRail franchise is provided by the Scottish Executive but the SRA retains responsibility for funding the Wales and Borders TOC. The Welsh Assembly Government can, however, allocate additional rail subsidy should it so wish. The element of subsidy provided to TOCs which ends up being paid to the network operator as track access charges could be spent anywhere in Great Britain.

26 Department of the Environment, Transport and the Regions (2000) *Transport 2010.*

27 HM Government (2002) *The government's response to the Transport, Local Government and the Regions Select Committee report, Passenger rail franchising and the future of railway infrastructure.* Cmnd 5472, The Stationery Office, London, 1.

28 Department of the Environment, Transport and the Regions (2000) *Transport 2010*, 6.7 and 6.8.

29 These measures do not have to involve new technology. For example, the reopening of certain mothballed signal boxes would increase the number of signalling blocks and, hence, enhance capacity.

30 There is clearly a trade-off here as many local stopping services are important in terms of social inclusion.

31 Richard Beeching was installed as the chairman of the British Railways Board in the early 1960s and charged with making the network profitable. He proposed large-scale route closures – many of which occurred – but was unable to prevent BR from returning losses; see British Railways Board (1963) *Reshaping Britain's railways. Part one – report and appendices; part two maps.* HMSO, London.

32 Although reinstating lines such as Oxford-Cambridge or the Waverley line from Edinburgh to Carlisle via Galashiels could bring major capacity benefits, particularly for freight operators.

33 Wolmar, C (2001) *Broken rails.*

34 Godward, E (1999) *Research and findings on the performance and delivery of a modern rail infrastructure by Railtrack.* Advanced Railway Research Centre, Sheffield.

35 Uff, J and Lord Cullen (2001) *The joint enquiry into train protection systems.* HSE Books, London; Uff, J (2000) *The Southall rail accident report.* HSE Books, London; Health and Safety Commission (2001) *The Ladbroke Grove rail inquiry, parts 1 and 2 report.* HSE Books, London; Wolmar, C (2001) *Broken rails.*

36 *The Daily Telegraph* (2001) 15 October.

37 Department of Transport, Local Government and the Regions (2001) *Concentrate on early gains for passengers: Byers spells out new franchising policy.* News Release 325, 16 July, 1.

38 Bradshaw, W (1991) *A review of policies for the future of Britain's railways.* Address to the Railway Study Association, 13 November, London.

39   Strategic Rail Authority (2001) *Freight strategy*. SRA, London.

40   Strategic Rail Authority (2002) *Strategic plan*. SRA, London.

41   *Rail* (2002) When is a government guarantee not a guarantee? When it's SRA support! 432, 4–5; *Rail* (2002) Network Rail's wait finally over as Railtrack deal is signed. 439, 4–5.

42   The Public Sector Borrowing Requirement is the net difference between the government's income and its expenditure.

43   Quoted in *Rail* (2002) Better passenger care and fewer trains may be helpful. 434, 3.

44   TAS Publications and Events Ltd (2001) *Railway finance monitor*. TAS Publications, Skipton; *Financial Times* (2002) 3 December.

45   The *Financial Times* alleges that a lot of public subsidy will be used to buy improvements through the private sector and therefore has been counted twice. A £50 million station upgrade, for example, may well have been factored in to the subsidy profile claimed by the winning TOC, so it gets counted as part of the government's £30.5 billion investment. Yet because it is the TOC who actually pays the contractor to effect the improvement, it counts as part of the private sector's anticipated £34 billion investment as well.

46   TAS Publications and Events Ltd (2002) *Rail industry monitor*. TAS Publications, Skipton.

47   The Observer (2002) 15 December. www.observer.co.uk/politics/story/0,6903,860336,00.html (accessed 15 December 2002).

48   House of Commons (2002) Session 2001– 2002, HC 558i, 22 March, 455.

49   TAS Publications and Events Ltd (2002) *Rail industry monitor*.

50   Commission for Integrated Transport (2001) *A comparative study of the environmental effects of rail and short-haul air travel*. http://www.cfit.gov.uk/reports/racomp/exec/index.htm (accessed 25 November 2002).

51   House of Commons (2002) Session 2001–2002, HC 558i, 22 March, 437. Traffic is projected to increase by 17 per cent if motoring costs fall by a fifth, but by 13 per cent if they remain constant.

52   Department of the Environment, Transport and the Regions (1997) *New roads as a last resort – Strang*. Press Release 216.

53   House of Commons (2002) Session 2001–2002, HC 558iii, 22 March, 120.

54   Department of Transport (1992) *New Opportunities for the railways*.

55   *The Economist* (1996) A great train cash. January 21, 20.

56   Friedman, M (1984) *Tyranny of the status quo*. Secker & Warburg, London.

57   Wolmar, C (2001) *Broken rails*.

58   Ford, R (2003) SRA exposes massive exploision in costs. *Modern Railways*, March 21–2.

59   Glaister, S (2001) *UK Transport Policy 1997–2001*; Affuso, L, Masson, J and Newbery, D (2000) *Comparing investments on new transport infrastructure: roads vs. railways?* DAE Working Paper 0021, Department of Applied Economics, University of Cambridge. See also *Financial Times* (2002) 26 July; *House of Commons Papers* (2002) Session 2001–2002, HC 558–I, 27 May.

6

Light Rail and the London Underground

Richard Knowles and Peter White

Urban railways play a major role in the London region, but a much more limited one elsewhere in the United Kingdom, even in major conurbations of over one million people. In this respect, parallels may be drawn between Paris and London, but a striking contrast exists between conurbations such as Greater Manchester and Stockholm or the West Midlands and Hamburg. There are no really comprehensive 'heavy' rail networks in cities outside London (although Glasgow is something of an exception), but rather a series of routes in the Passenger Transport Executive (PTE) areas operated by British Rail's (BR's) successor companies whose pattern has been determined by history, rather than recent planning or investment (Chapter 5). The exceptions comprise mainly the light rapid transit (LRT) systems (including the Tyne and Wear Metro), which are the result of more recent planning and investment decisions. Even these, however, currently serve only selected corridors rather than provide extensive networks. Data on walking time access to stations in the National Travel Survey illustrates the differences. In London, 61 per cent of the population is within 13 minutes' walk of a rail station (underground and/or surface rail), whereas in PTE areas this falls to about 21 per cent.[1]

In terms of passengers carried, London Underground accounts for almost 1,000 million passenger trips per year, equal to the entire 'National Rail' network operated by BR's successor companies (albeit with a much shorter average trip length). The next three largest systems – Merseyrail, the Strathclyde PTE/ScotRail Glasgow area network, and Tyne and Wear Metro – each carry between 30–45 million passengers per year. This is not merely due to a lower population being served, but a much lower rail trip rate per head. Other networks carry about 15–25 million per year (West Yorkshire, West Midlands) or very small numbers (Tyne and Wear non-Metro services, South Yorkshire). Greater Manchester now carries only about 14 million per year, but this is largely explained by the two busiest routes being transferred to the Metrolink light rail operation.[2] Another system is the Glasgow Underground, a circular route serving the central area and western inner city, operated by the PTE, carrying about 15 million passengers per year.

Until the 1970s, urban railways in Britain could be seen as suffering from a similar decline to that faced by buses, as car ownership rose and employment declined in city centres. Suburban services in cities such as Bristol and Edinburgh virtually disappeared in the 1960s. Growing congestion led to an increased interest in rail's potential, however. This resulted in some upgrading and capacity increases, including new underground route sections on the Glasgow and Merseyside PTE networks. Unfortunately, continued urban economic decline meant that usage was modest.[3] Subsequent growth in PTE rail use has perhaps been associated more closely with increased central area activity in other cities, notably in the West Yorkshire services focusing on Leeds. Since rail privatization, overall use has grown rapidly, but urban networks outside London have made only a small contribution to this. Conversely, growth on the London and South East group of Train Operating Companies (TOCs) – whose reliability was (ironically) less affected by the Hatfield crash impacts than regional or long-distance services (Figure 5.4) – has continued, representing over half of all the national increase since 1995/96.[4] Light rail systems have also grown substantially since 1992 – they form the dominant rail provision in Greater Manchester, South Yorkshire and Tyne and Wear – although experience has varied widely between networks.

This chapter examines the principal urban rail systems in the UK beyond the National Rail network (Chapter 5). It begins with a review of recent developments in LRT and assesses the government's apparently changing policy towards the mode. The London Underground is the subject of the analysis in the second section of the chapter.

Light Rail

Light rail, or LRT, provides a small but increasing contribution to public transport in the UK. It has reemerged recently to play an important part in the government's integrated transport policy for conurbations and large cities, after falling out of favour with successive Conservative and Labour administrations. LRT provides an attractive and sustainable alternative to the private car. Like all good quality public transport, it reduces social exclusion by making access easier for noncar owners to work, learning, health care, shopping and recreation and leisure facilities.[5] Its proven ability to secure modal shift from cars helps to limit urban traffic congestion and reduce air pollution from motor vehicle emissions without occupying road space – and thus alienating car drivers – apart from along a few city-centre sections of some route networks.[6] As LRT systems are electrically powered, they are pollution-free at the point of use and potentially capable of using energy from renewable sources. Although fixed track light rail systems have inflexible routes and are high-cost compared with buses, their visibility and positive image help them attract passengers and achieve modal shift more easily.[7] Light rail is therefore a more effective catalyst for economic development and urban regeneration than buses and is currently playing a key role in the renaissance of large cities such as Manchester.[8] Indeed, David Begg, Chair of the Commission for Integrated Transport (CfIT), sees light rail as a statement that a city is committed to regeneration.[9]

Whilst bus services will continue to provide most local public transport outside rail-dominated central and inner London, light rail has major advantages of larger capacity, higher speeds and greater reliability, which are particularly valuable in major intraurban corridors of up to 12 miles in length. Light Rail's average speeds of 12 to 25 miles per hour (mph) and capacity of up to 20,000 passengers per hour compare favourably with 6 to 12 mph and up to 5,000 passengers per hour by bus.[10] Light rail systems provide a cut-price alternative to underground, heavy rail metros because they utilize converted railway lines, largely segregated road alignments, short sections of tunnel or a mixture of all three. Like heavy rail systems, however, LRT projects have a long lead time (typically 8 to 10 years from conception to delivery). Although this gestation period makes LRT a 'slow win' and, therefore, potentially less attractive than buses to politicians seeking quick solutions to transport problems, strong opposition from the motorists' lobby to bus priority schemes – which reallocate road space away from cars – has led to a reassessment of light rail as a 'win-win' solution to traffic congestion on some major urban corridors. LRT investment can enable politicians to be portrayed as tackling traffic congestion without upsetting the politically powerful road lobby.

Light rail development under the Conservatives

Light rail first emerged in the UK in the 1970s and 1980s. In many European cities, particularly in Germany, tram or streetcar systems have been progressively upgraded to segregated light rail avoiding congested roads by putting lines underground in and near city centres.[11] In contrast, for over 30 years from 1946, British governments short-sightedly advised local councils to regard trams as an outdated mode of transport, and towns and cities across the UK had closed down their systems by 1962. Only Blackpool's seafront tourist line remained in service.

Before bus deregulation outside Greater London in October 1986 (Chapter 7), light rail was conceived as the backbone of integrated urban transport systems in the provincial conurbation centres of Newcastle, Manchester and Sheffield.[12] Of these three LRT schemes, only Tyne and Wear Metro, the first line of which opened in 1980, was operational in time to be able to demonstrate an integrative role, with many parallel bus routes shortened to feed into and distribute from the light rail network (Table 6.1). This integration collapsed with bus deregulation and the Metro's patronage fell sharply from a high of 61.1 million passengers in 1984/85 to 46.4 million in 1986/87 and then gradually to 32.5 million in 2000/01. The decline was due to direct bus competition and higher car ownership and use, although light rail usage in the area has recently started to increase.

Britain's second light rail system was a new-build, largely elevated route with driverless trains which opened in 1987. The system was not, in fact, in Manchester or Sheffield, but in London's Docklands, and it formed part of an integrated transport system made possible because London's buses and Underground were not deregulated. The Docklands Light Railway (DLR) linked derelict dock land in the Isle of Dogs with Tower Gateway, in the City of London, and Stratford. The DLR has proven to be a successful catalyst for the redevelopment of derelict docks for offices, housing and shops. It has been extended three times and is now Britain's busiest LRT system with 41.3m passengers in 2001/02 (Table 6.1).[13] This success has been achieved despite a 1989 government report which argued that rail investment would have little impact on London's traffic congestion.[14]

A change in government policy required new LRT systems outside London to be privately operated without revenue subsidy, as a condition of a maximum 50 per cent capital grant from Westminster towards the cost of construction. Tougher government guidelines introduced in 1989 also required nonuser benefits – including urban regeneration, fewer accidents, lower road traffic congestion, less noise pollution and lower vehicle emissions – to exceed the value of the capital grant. Manchester Metrolink's Phase 1 conversion of suburban railway lines from Bury and Altrincham, linked on-street through Manchester city centre, and Sheffield Supertram's three new-build routes, two of them mainly on-street, overcame these difficult hurdles and began construction, opening in 1992 and 1994/95 respectively (Figure 6.1). Demonstrating further the effect of bus privatization and deregulation on the plight of LRT, a planned scheme in Strathclyde was abandoned after a legal challenge from local bus companies.

Despite this harsh policy environment, a boost came in 1991 through a House of Commons report, *Urban Public Transport, the Light Rail Option*, that strongly recommended LRT and suggested improved evaluation, approval and funding procedures. Although none of the Committee's recommendations were implemented by the Conservatives, at the time more than

Table 6.1 English light rail systems

System	Line	Opened	Route (km)	Cost (£m)	Cost (£/km)	Passengers (millions) 2000/01	2001/02
Tyne & Wear Metro	Tyneside Loop, Heworth & S Tyne	1980/4	55.0	284	5.2		
	Airport	1991	3.5	12	3.4		
	Sunderland	2002	19.2	56	2.9		
	Total		77.7	352	4.5	32.5[1]	33.4[2]
Docklands Light Railway (DLR)	Tower, Isle of Dogs & Stratford	1987	12.0	77	6.4		
	Bank	1991	1.5	282	188.0		
	Beckton	1994	8.0	258	31.5		
	Lewisham	1999	4.5	220	48.9		
	Other work	various			117.0		
	Total		26.0	954	36.7	38.4	41.3
Manchester Metrolink	Altrincham–Bury	1992	30.9	149	4.8		
	Salford Quays & Eccles	1999/2000	7.5	120	16.0		
	Total		38.4	269	7.0	17.2	18.2
South Yorkshire Supertram (Sheffield)	Meadowhall, Hillsbro. & Halfway	1994/5	29.0	260	9.0	11.1	11.4
Midland Metro	Birmingham–Wolverhampton	1999	20.4	145	7.1	5.4	4.8
Croydon Tramlink	Phase 1	2000	28.0	200	7.1	15.0	18.2
Nottingham Express Transit	Transit Phase 1	2003	14.0	220	15.7	–[1]	–[2]
TOTAL LIGHT RAIL			233.5	2400	10.3	119.6[1]	127.3[2]

Sources: Department for Transport (2002) *Transport statistics Great Britain*. DfT, London; Knowles, R (1996) Transport impacts of Greater Manchester's Metrolink light rail system. *Journal of Transport Geography*, 4, 1–14; London Transport (1997) The Docklands Light Railway story. In London Transport (ed.) *Starting from scratch*. London Transport, London, 44–83; Young, A (2001) *10 years of success with Manchester Metrolink – experience for other cities*. Paper presented at Light Rail 2001 conference, Manchester, November.

Notes

1 Sunderland and Nottingham lines not open.
2 Sunderland line opened March 2002, Nottingham not open.

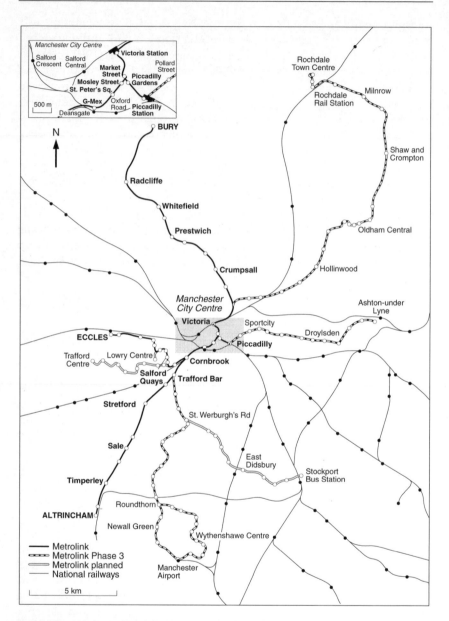

Figure 6.1   Manchester Metrolink: existing network, Phase 3 and future plans.

40 light rail systems were being built, planned or considered in Britain.[15] Many local councils, deprived of control of their local bus networks by bus deregulation and privatization, were attracted to light rail as a public

transport mode they could control and own even if its operation was franchised. The Tyne and Wear Metro's extension to a park and ride station at Callerton Parkway and Newcastle Airport opened in 1991; Acts of Parliament were obtained for West Midlands Phase 1 and Avon light rail schemes; and Parliamentary Bills were submitted for light rail routes in Croydon, Leeds and Nottingham. Legal powers for LRT routes could also now be acquired more quickly and cheaply through the use of Transport and Works Act Orders, rather than the previous requirement, dating back to the turnpike, canal and steam railway eras, of an individual Act of Parliament for any new public transport route.

The future for light rail looked bright, especially after Manchester Metrolink Phase 1 exceeded its patronage targets and secured an unexpected modal shift of 2.6 million former car users, representing 22 per cent of its passengers.[16] Furthermore, the volume of road traffic on the main radial routes running parallel to Metrolink into Manchester city centre fell by up to 10 per cent. Sheffield Supertram also secured a 22 per cent modal shift from cars to light rail.[17] The high expectations of a golden era for light rail were, however, curtailed by cuts in government expenditure after the 1992 general election, due to the economic recession.[18] Funding for LRT schemes was reduced, and there was a four year gap after 1995 before Midland Metro opened from Birmingham to Wolverhampton, Metrolink was extended from Manchester to Salford Quays and the DLR was extended under the River Thames to Lewisham in south London (Figure 6.1, Table 6.1). The DLR extension was followed in 2000 with the completion of Metrolink Phase 2 from Salford Quays to Eccles and the opening of Croydon Tramlink.

A new deal for light rail?

Perhaps surprisingly, the 1997 Labour government's much vaunted integrated transport policy proved to be bad news for light rail schemes, at least at first. The long-delayed White Paper, *A New Deal for Transport: Better for Everyone*, prioritized investment in buses over light rail and unexpectedly perpetuated the Conservatives' policy of bus deregulation, which had failed to arrest a long-term decline in bus use (Chapter 7).[19] Whilst *A New Deal for Transport* claimed that light rail was a part of the strategy for integrated urban transport, it noted that LRT's capital cost was high and that bus schemes could prove more cost-effective and spread the investment benefits more widely. The government saw statutory bus Quality Partnerships between local councils, who as highway authorities own the road space, and the mainly privately owned bus companies as the principal way of improving urban public transport and securing modal shift from cars (Chapter 7).

This was despite the fact that most Quality Partnerships – which did exist before the Transport Act (2000) but without the support of a central government framework – had failed to promote modal shift.[20] *A New Deal for Transport* made clear that LRT projects would not be regarded as priority for funding and must demonstrate good value for money and form 'an integral and necessary part of a Local Transport Plan' (Chapter 3).[21] In other words, light rail was too expensive and took too long to construct to be politically attractive. Not only were Quality Partnerships seen as cheap and capable of delivering the impression of public transport improvement quickly, but, because the Partnerships were initiated by local councils, it was these authorities, rather than the government, who would be blamed by motorists angry at the loss of road space to bus lanes.

Labour ministers further downgraded light rail by implying that they expected the construction of major new LRT schemes to wait until revenues could be drawn from road user and workplace parking charges (Chapter 4). This again left local authorities to face the wrath of the car-owning electorate. Although the Mayor of London is introduced a £5 daily road user charge for central London in February 2003, most other local authorities are waiting to see its effects before deciding whether to follow suit.[22] Will the revenue raised exceed losses from diversion of jobs and economic activity to outside the charging zone and how strong will the local political opposition be? It is currently estimated that, outside of London, only three cities will introduce road user charging and one will set up a workplace parking levy by 2010 (the government's targets had been 8 and 12 respectively) (Chapter 4).[23] This has obvious implications for the funding of LRT schemes across England.

The government's hard-line stance on light rail appeared to soften in December 1999 when the then Secretary of State for Transport, John Prescott, predicted that we would see more light rail systems, giving people a modern, attractive alternative to the car. Although expectations were dampened again when the government restated the view that bus priority measures and guided bus schemes may be more cost-effective than LRT, the potential for a light rail renaissance dawned unexpectedly in the summer of 2000. Perhaps influential in precipitating this was the strong vote of confidence in LRT voiced by the House of Commons' Environment, Transport and Regional Affairs Select Committee. It argued that 'people will not switch to public transport unless it is reliable, frequent, efficient, safe and clean with affordable fares. Light rapid transit systems meet these criteria, and so, where appropriate they should be pursued'.[24] The Committee recommended a raft of measures in support of its position (Table 6.2).

Table 6.2   Recommendations of the Environment, Transport and Regional Affairs Select Committee

- Government investment should be directed to the mode of transport best suited to local conditions and passenger flows whether heavy rail, light rail or buses
- A strategy of investment in complete light rail systems, rather than isolated lines, should be made because a 'network effect' would generate extra passengers
- Light rail vehicles should be standardised to reduce costs compared with the existing practice of different specifications for each light rail system
- Light rail should be given priority over other vehicles when running 'on-street' to minimise disruption
- Complementary car restraint measures should be implemented when new LRT schemes commence operation to increase the amount of modal shift
- If Quality Partnerships prove ineffective in integrating light rail and bus services, they should be replaced by stronger Quality Contracts
- The government should amend planning guidance to further emphasise the need to integrate LRT schemes with urban development
- The Strategic Rail Authority (SRA) should examine the potential for converting heavy rail routes to light rail and for track-sharing as part of re-franchising of passenger rail services

Source: House of Commons (2000) Session 1999–2000, HC 153, 8 June. The Stationery Office, London.

In July 2000, the government unveiled the detail of its transport strategy for England as part of *Transport 2010: The 10-Year Plan*. This included both a target to more than double light rail use in the country by 2010, and an *aspiration* to deliver up to 25 new light rail routes. This aspiration was intended to reassure many cities and conurbations that their wish to develop light rail schemes might be supported and part-funded in future by central government instead of being obstructed and rejected. The government's more upbeat assessment of light rail's role echoed the sentiments of the Committee, in stating that:

> Light rail, trams and other rapid transit systems can play a significant part in improving the attractiveness and quality of public transport in major conurbations. They can move large flows of passengers quickly and reliably. They compete with the car in terms of journey times and convenience. And they help to reduce congestion and pollution.[25]

Ministers' commitment was now to fund a substantial increase in the role of light rail in Britain's larger cities and conurbations over the next few years, supporting schemes that they regarded as offering good value for money as part of integrated transport strategies. Funding would come through public/private partnerships between central government, local councils or Passenger Transport Authorities (PTAs), developers and franchised light rail service operators. Revenue from local congestion charging systems was assumed to contribute to this financial package although, of course, it is unlikely this will materialize in most large cities over the next 10 years. The Conservatives' guidelines that Westminster's contribution to the construction costs must not exceed 50 per cent and that nonuser benefits must exceed the value of the grant no longer pertain, although a local contribution is required.

In its subsequent formal response to the Committee's report, the government endorsed MPs' support for LRT systems where they provide the best solution to a transport problem.[26] Although such a form of expression offered a potential 'get-out' clause for ministers – all manner of reasons could have been used to justify a decision not to fund various LRT schemes when proposals were advanced – ministers committed funds towards the cost of four light rail schemes: the Tyne and Wear Metro's £100 million extension to Sunderland (which opened in 2002); Nottingham Express Transit (due to open in November 2003); the DLR extension to London City Airport; and Phase 3 of Manchester Metrolink. This last scheme is a three-route, 35-mile extension and forms Britain's most extensive single light rail contract. It includes a 15-mile route to Oldham and Rochdale using converted suburban railway line and new on-street town centre routes, together with new-build routes to South Manchester and Manchester Airport (13.75 miles), and Ashton-under-Lyne via City of Manchester Stadium (6.25 miles) (Figure 6.1).[27] The selection of a contractor in 2002 to design, build, operate and maintain Metrolink Phase 3, at an estimated cost of £527 million, and take over operation and maintenance of Phases 1 and 2 on a 15 year franchise, was delayed by a reported £200m funding shortfall due to the higher contract price and/or smaller developer premium than previously anticipated.[28] The government confirmed funding approval in December 2002.[29]

Whilst the government's altered stance on LRT is to be welcomed, it is important not to overstate the degree of support which is on offer. The *10-Year Plan*'s target of 'more than doubling' light rail use by 2010 is both imprecise – the document failed to specify baseline or target figures – and undemanding.[30] Light rail use totalled 93.9 million passenger journeys in 1999/2000, so a 10-year target of at least 188 million passengers per year might be assumed. When the Transport, Local Government and the Regions Select Committee (the successor to the Environment, Transport and Regional Affairs Committee) challenged the government about the start date for the *10-Year Plan*, however, its response was that the *Plan* provides funding for 10 years from 2001. This implies that the baseline figure for light rail is 119.6 million passengers in 2000/01 with a revised target of at least 240 million passengers per year. In sustainable transport terms, this higher target is clearly a step in the right direction, but it is still rather less than challenging. Most of the extra passengers will come from opening already approved LRT systems – a study for CfIT suggests that the government's targets require only half of the 25 potential new light rail routes to open – and patronage growth on existing ones.[31] A 23 per cent passenger increase had already been achieved in just the one year between 1999/2000 and 2000/01 due to the Croydon Tramlink opening and growth in DLR and Metrolink traffic.

The *10-Year Plan* in fact fails to set specific targets for increasing the use of existing LRT systems, which seems a missed opportunity since substantial additional traffic growth and modal shift could be achieved by better marketing of existing light rail systems and through modal integration via Bus Quality Partnerships or Contracts. This latter form of reregulation would allow buses to feed into and distribute from light rail routes, without the threat of direct competition from other bus companies, as was achieved so successfully by Tyne and Wear's Metro and integrated bus services before bus deregulation in 1986. Whether Labour decides to pursue Quality Contracts in any meaningful sense remains to be seen, however (Chapter 7).

There are also no targets for the amount of investment in light rail – although, of course, this could be misleading if money were wasted – or the length of new routes to be opened. The *10-Year Plan* assumes about £2.6 billion of public and private investment in light rail, but CfIT saw considerable uncertainty about the availability of private sector funding for rail projects – including LRT – and warned that this could act as a potential barrier to the achievement of the *10-Year Plan*'s targets. CfIT's analysis was informed by the inability of many of the franchised passenger rail-operating companies to operate profitably, the government forcing the track and stations owner Railtrack into administration, and the failure of the mainly privately funded Metrolink Phase 2 to reach its passenger targets (Chapter 5).

The Department for Transport (DfT) confused the issue of targets in summer 2002 by representing the national 10-year targets for light rail and bus at a combined growth figure of 12 per cent by 2010 over 2000 levels. This in effect removed any growth targets for both modes' traffic outside London, since the increase in the capital's bus passengers alone is likely to deliver the new growth objective (Chapter 7).

Despite the ambivalence of successive governments towards LRT and the undemanding targets and vague aspirations contained within the *10-Year Plan*, it does seem, rather surprisingly, that England could be on the verge of becoming a European leader in light rail.[32] Admittedly this is partly because countries like France, Germany, Italy and Spain – all of which are significantly expanding their rail-based urban transport networks – are investing huge sums in building full metros instead of, or as well as, focusing on LRT. Nevertheless, at present only Germany has a greater light rail route length than England (well over 125 miles of track is now in service in the latter), and most major English cities now either have light rail or will acquire it soon (Table 6.1). In addition to the four new schemes mentioned above, Leeds obtained government funding and Private Finance Initiative (PFI) credits in 2001 for a three-route, 17.5 mile Supertram light rail network costing £500 million. The government is also backing a £190 million system in South Hampshire and has approved in principle plans to develop Merseytram, a major light rail line in Liverpool. DLR, Metrolink, Midlands Metro and South Yorkshire Supertram have further expansion plans. Other towns and cities across the UK (including Belfast, Blackpool, Bristol, Cambridge, Cardiff, Glasgow, Hull, Leicester and Middlesbrough) are developing light rail proposals, but only a very limited number are likely to open within the next 10 years.

Light rail's future in the UK looks bright in comparison with its recent past. LRT's ability to achieve modal shift without completely alienating the powerful road lobby, its key role in urban regeneration and its high-profile image makes light rail attractive to politicians who want to be seen to be tackling traffic congestion and improving public transport. The government's changing attitude to LRT is to be welcomed, especially since other aspects of its transport policy seem to be backing away from the sustainable transport agenda. Yet LRT can play a far greater role in a policy of integrated transport than ministers currently envisage. The *10-Year Plan's* light rail traffic targets need to be much more ambitious, with separate targets to increase traffic on each existing system as well as meeting traffic forecasts on new systems. Achieving such new and revised targets might require the further pursuit of demand management measures and would almost certainly demand the introduction of Quality Contracts to integrate bus services with light rail networks and operations (Chapter 7). The latter is not such a huge change given the near monopoly control of most large urban bus markets by single privatized bus companies, but would demonstrate a genuine commitment to the aim of developing a more sustainable transport system.

The London Underground

Until the mid 1980s, the role of rail in London was declining. Underground use had fallen from a peak of 702 million passengers in 1951, as central area employment declined and fares rose in real terms.[33] Following the completion of the Victoria Line (Walthamstow–Victoria) with its Brixton extension in 1971, no new routes were constructed until the Baker Street – Charing Cross route opened in 1979 which linked with the Baker Street – Stanmore section of the Bakerloo Line to form the Jubilee Line. While renewal of the rolling stock fleet continued, there was no perceived need to raise peak capacity. Ridership reached a low point in the early 1980s (498 million in 1982 – 29 per cent below the 1951 level), following the sharp fare increases after the 'Fares Fair' Lords' judgment.[34]

Subsequent events have provided a remarkable turnaround in ridership trends, although not necessarily in all performance characteristics (Table 6.3). During the mid-to-late 1980s ridership rose sharply, to reach 815 million in 1988/89. This was associated with two main factors. The first was the introduction of the bus/rail 'Travelcard' in 1983. This encouraged a shift from bus and car for peak radial commuting journeys, especially by removing the price penalty of interchange with bus. It also stimulated growth in off-peak travel, additional journeys effectively being free of monetary cost to those who had already purchased their card for peak travel. Hence, the percentage growth at off-peak times was higher than for the market as a whole. The second factor was a reversal of the decline in central London employment, associated with growing activity in sectors such as financial services and tourism.[35]

There was a decline in peak traffic in the early 1990s, due to a fall in central area employment (a function of national recession rather than a shift to other areas), but this reversed once more from the mid-1990s. Further marketing efforts, such as the family and weekend Travelcards, have helped to stimulate off-peak traffic. Total ridership thus grew to an all-time high of 970 million in 2000/01. To this may be added the traffic of the DLR of 38 million in that year.

Table 6.3 London Underground patronage, 1990/91–2000/01

	Passenger kms (millions)	Passenger journeys (millions)
1990/91	6,164	775
1991/92	5,895	751
1992/93	5,758	728
1993/94	5,814	735
1994/95	6,051	764
1995/96	6,337	784
1996/97	6,153	772
1997/98	6,479	832
1998/99	6,716	866
1999/2000	7,171	927
2000/01	7,470	970

Source: Transport for London (2002) *Transport statistics for London 2001*. TfL, London.

Some efficiency gains, coupled with traffic growth and real fares increases, created a situation in which operating losses ceased, and a modest surplus over operating costs was attained, contributing to capital renewal needs. In this respect, London is virtually unique among metro systems in North America and Western Europe, where operating support is required as well as capital investment from public funds. Other examples, such as Hong Kong, are in cities of exceptionally high activity densities, which are reflected in ridership densities per route kilometre.[36] Contrary to the impression which many Underground users may have, London is in fact a system with a lower ridership density (passenger km per route km) than the worldwide average for metro systems, albeit this figure stems in part from extensive low-density outer sections (such as the Central Line to Epping), and the slightly lower capacity per car of 'tube' stock than that of other subsurface metros.[37]

Growth in central London employment stimulated the development of Docklands, especially the financial and business services district at Canary Wharf. Serving this development was also the main objective of the Jubilee Line Extension (JLE) agreed by the Conservative government in the early 1990s. Running from the existing Jubilee Line at Green Park, via Canary Wharf to Stratford, this route has been constructed at very high cost and largely from public funds, despite a small developer contribution. Construction was delayed, and as a result it opened in 1999, after the present government came to power.[38]

The Underground under Labour

Following transfer of London Transport from the then Greater London Council in 1986 to a board appointed by central government (London Regional Transport), integrated control and ownership of the Underground and buses continued until the early 1990s, when ownership of the Underground was invested in a subsidiary company, London Underground Limited (LUL). The bus companies were then privatized, but in contrast to the rest of Britain an integrated network planning and pricing structure has been retained, since bus services are operated under contract rather than in the deregulated framework found elsewhere (Chapter 7). Bus network restructuring continued to create improvements in suburban areas, with services often acting as feeders to underground stations. This approach has also contributed to the success of the Croydon Tramlink.

The first major change introduced by Labour to affect the Underground was the creation of the Greater London Authority (GLA) together with an elected Mayor under the Greater London Authority Act (1999). Both came into power in 2000. Bus responsibilities were transferred to the GLA, which manages its transport role through a new agency, Transport for London (TfL). This now controls buses, major roads, river services and taxis, permitting a high degree of policy integration. Major policies, such as fares levels, are effectively controlled by the Mayor. Bus use has been stimulated by lower fares and expanded services. The Underground is due to be transferred to TfL now that issues relating to its Public Private Partnership (PPP) funding package have been resolved (see below).

As evident from the previous discussion, the situation inherited by the incoming Labour government in 1997 displayed some favourable features in terms of ridership trends and financial performance. However, system capacity, both in terms of rolling stock and infrastructure, was proving inadequate to cater for growing peak demand. Since in practice only a very small part of the JLE cost was met by private sector contributions, the available public funding for investment in the Underground was largely directed toward the JLE rather than the renewal of the existing network. In addition, much of the infrastructure is now very old and requires major work to renew existing installations, quite apart from creating additional capacity (this includes signalling, power supply and, in some cases, tunnels and other fixed structures).

It has already been mentioned that the Underground was able to fund at least some of its necessary renewal investment from its operating surplus. This surplus grew to a peak of £265m (at current prices) in 1997/98, but fell to £73 million in 2000/01. In the following year it became a loss of £120 million.[39] The worsening in 2001/02 was primarily due to a growth in labour costs, both in terms of the number of staff employed and real wage levels.[40] It is important to note that these figures are before depreciation.[41] Since in 2000/01 this was £340.8 million, only a modest contribution was made to covering it. London Underground's situation may be compared with that in the bus industry, where an aggregate operating surplus has been made after historic depreciation. This is generally sufficient to meet the difference between historical and replacement depreciation, thus enabling steady fleet renewal.[42]

Investment in the system has been far higher than the operating surplus, because of both the scale of renewal needed in the existing system, and the opening of new routes such as the JLE. Between 1994/95 and 1999/2000 inclusive, for example, 53 per cent of the total investment expenditure was devoted to the JLE. At 2001/02 prices, the overall volume of investment peaked in 1995/96 at £1260m, falling to £464m in 2000/01, although recovering to £618m in 2001/02.[43] Most investment requirements have been met by grants through public expenditure, apart from the element of developer contribution to the JLE. This budgeting has generally been planned on a short time-scale (one year, in some cases with further short-term changes), making it difficult to programme work efficiently. As in other sectors of expenditure, however, Labour has been attracted by the concept of the Private Finance Initiative (PFI), which has evolved into the PPP. Under this arrangement, private companies take on a package of investment and maintenance, being funded by a series of annual payments. The public sector avoids the need for 'up front' initial investment, and outward cash flows in the first few years of such projects will be lower in consequence. The private provider is incentivized to make a good job of the new work, so that future maintenance costs are minimized.

Several examples of this approach may already be found on the Underground, including the 'Prestige' smartcard ticketing project, the provision of new rolling stock for the Northern Line in 1998-2000, and the renewal of power supply. The cost of these initiatives appear as an annual outflow on LUL's balance sheet: in 2000/01, for example, 'leasing and PFI charges' incurred a cost of £189.7 million.[44] For several years ministers have proposed applying this concept on a far greater scale, to cover the renewal of assets on the system as a whole, including tunnels, power supply and rolling stock. In preparation for the introduction of the PPP, LUL has already been divided internally between three infrastructure companies (infracos) and the passenger train operating business.

The extent to which the public sector can make a saving over the whole period of such a contract is crucially dependent upon the interest rates (and hence applicable discount rates for evaluation purposes) faced by central government and the private sector.[45] At low interest (and discount) rates, a greater weighting would be given to future cash flows than at high rates. Hence, whereas in seeking to justify an investment project, low interest and discount rates are generally helpful (by increasing the value of discounted benefits vis-à-vis initial capital cost), correspondingly low rates tend to make a PFI/PPP scheme less favourable, by giving greater weight to the outward cash flows in later years. In general, the public sector can raise capital at far lower interest rates than the private sector. Hence, at the same efficiency level, undertaking work by the PFI/PPP route will generally cost *more* to the public sector in the long run than direct funding. In the Underground PPP scheme, private sector consortia winning contracts to take over the Infracos – the train operating business will remain under public control – will be paid in the form of an annual charge. Part of this will be met, in effect, by transferring LUL's existing budget for maintenance and renewal work, but it is clear that a substantial public sector contribution will also be needed, especially in the first few years of the contracts.

These issues were examined in a study of the Underground PPP for the Mayor of London by the Industrial Society in 2000 to which one of the authors acted as a specialist adviser.[46] The study highlighted a number of issues, notably the extent to which risks could be transferred wholly to the private sector, or would remain with a public body.[47] For example, while construction cost risk could be transferred, there is uncertainty about the condition of current assets and hence the true extent of remedial work required.

Analysis in the study by Nick Bunker also provided a useful illustration of the alternative costs of PPP and public funding. Assuming the same evaluation period (30 years), fares constant in money terms and some efficiency gains under both private and public sector funding through long-term programming of work, the PPP option would only be cheaper if the private sector companies could undertake a given quantity of renewal and investment work for about 20 per cent less than under direct public funding. This was a consequence of higher cost of capital for the private sector, especially for equity capital at around 15 per cent. Such large cost reductions must appear questionable, especially given the uncertainty due to conditions of the assets being taken over. What is more, matters have worsened due to the delays associated with a dispute over the introduction of the PPP – which resulted in an extensive legal dispute between the Mayor/TfL, who are vigorously opposed to the scheme, and central government – during which the investment backlog has grown. Contracts with preferred bidders were signed in August 2002 and the PPP is in the process of implementation. A final agreement was signed with the Tube Lines consortium for the Jubilee, Northern and Piccadilly infraco on 31 December 2002. A subsequent statement by Tube Lines' Chief Executive indicated that the company expected to make a 19 per cent return on equity, or an overall margin for the first seven and a half year period of its contract of 11 per cent after tax. The consortium will receive an infrastructure service payment of £300 million per annum. From 4 April 2003 the contract came into effect with the Metronet consortium for the other two infracos, Bakerloo Central Victoria and Sub Surface Lines.

The Mayor and TfL – along with others, such as Tony Travers – have proposed that, instead of the PPP, bonds be issued to finance the investment. Private contractors would still be used to carry out the work itself. Such an approach has been opposed by central government on the grounds that bonds would count against the Public Sector Borrowing Requirement (PSBR).[48] An inconsistency is evident in Labour's policy, in that state guaranteed bonds have been issued for phase two of the Channel Tunnel Rail Link (CTRL). In addition, similar bonds to support Network Rail (the successor to Railtrack) have been declared by the Office for National Statistics (ONS) not to form part of public borrowing, although the National Audit Office (NAO) appears to differ in its interpretation (Chapter 5). Clearly what matters in the long run is the *real* cost of undertaking the work, not the notional classification of capital borrowings as 'public' or 'private'.

It is not entirely apparent why the government is so strongly wedded to the PPP concept, especially since its predecessor – the PFI – originated under the previous Conservative government. None the less, if the only alternatives are PPP or continued very low investment levels funded directly by government, the PPP would be preferable as a means of ensuring that renewal investment does take place, albeit not necessarily at best value for money in the long run.

The Underground and sustainability

The question of sustainability, when applied to the Underground, has two aspects. Firstly, can the existing system be sustained in the sense of renewing its assets to maintain at least the existing capacity and service quality? As indicated above, this is largely dependent upon the means by which investment can be funded (although it does not appear practicable to sustain the system wholly from passenger revenues).

Secondly, in what sense does the Underground contribute to wider sustainability objectives? Given the continued existence of a large city such as London, it clearly forms an essential element in the transport system. There is insufficient capacity to accommodate demand on the road network, especially for peak central area commuting. Moreover, high-density cities with a significant dependence on public transport consume a substantially smaller share of GDP to provide transport for their citizens, and also less energy per capita, than low-density cities such as those of North America or Australia.[49] For example, journeys made within London and Paris incur a cost to the community equivalent to about 7 per cent of GDP, compared with 12 per cent in Melbourne and 14 per cent in Houston. Annual energy consumption (Megajoules per inhabitant) in London and Paris is around 15,000 compared with 32,000 in Melbourne and 86,000 in Houston.[50] While associated with average urban density, these differences may also be due to factors such as real fuel prices, as the late Mike Breheny has pointed out.[51]

Some of the additional off-peak traffic stimulated by marketing measures might be seen as adding to the use of resources, vis-à-vis more local trips than might otherwise be made. As most of the Underground's costs are fixed or determined by peak demand, however, the marginal cost of additional off-peak services is very low. Even in energy terms, the Underground's use of electricity gives greater scope for alternative primary sources, together with potential not yet fully exploited for regenerative braking, and combined heat and power thermal stations.

A notable contrast exists between policy for the Underground and national railways, in that Underground prices have generally risen in line with inflation (both under central government, and the Mayor's, policies) whereas since 1999 the annual increase in season ticket prices on the franchised TOCs has been restricted to one per cent below inflation. Hence, the real cost of long-distance commuting has fallen by about 6 per cent since 1999, and a stimulus is given to greater energy consumption as a result.[52] This policy also creates problems in maintaining an integrated Travelcard pricing structure. TOCs are incentivized to aim for the long-distance peak market, since falling real revenue per passenger km increases the break-even distance at which additional peak traffic may be worth attracting.

Within London, one might wish to divert movement from congested road networks onto TOCs' services, yet this would be unattractive to the operators. Even the cost of infrastructure improvements and the additional rolling stock necessary to raise capacity simply by lengthening existing trains is unlikely to be covered by extra revenue. Expanding the Underground network is one means of improving access in inner London, yet this might be attained far more cost-effectively by new infrastructure and services on surface lines. For example, the JLE, in addition to serving Canary Wharf, also provides new stations in parts of London poorly-served before, such as Bermondsey.[53] Such low-income areas thus potentially benefit from large journey time savings to the central area, although the cost of a new deep-level station is in the order of £100 million. At a similar distance from central London is the high-density area around Camberwell Green, close to the Thameslink route. A station here could be built on the surface at a more modest cost and, indeed, is now being assessed by the SRA, but the exact pattern of services that would call there is unclear.[54]

Likely Future Developments

Continued growth of population and associated economic activity in London is likely to stimulate further growth in demand for the Underground. Under the PPP, some increase in capacity on existing lines will be attained, as resignalling and higher-performance rolling stock are introduced. However, this will only be in the order of 1.0–1.5 per cent annually, simply coping with likely demand growth rather than reducing overcrowding. Now progressing (albeit slowly) is the East London Line Extension project, lengthening the existing north-south line located to the east of the central area (New Cross–Whitechapel) north and south over existing surface rail routes.[55] Major new schemes are proposed, notably the Hackney–South West line, but the very high cost of new tunnelling in the central area is a serious constraint.

The transfer of the Underground to control of the Mayor and TfL in 2003 following implementation of PPP will permit integrated management of the public transport systems, but will potentially lead to conflicts between the

Mayor and the existing management, given his negative public comments on its performance. One must also still question whether the type of vertical separation which has been introduced in London is necessarily appropriate. If ministers judge that some form of private sector involvement in the running of the Underground is absolutely necessary, a case can be made for leaving the infrastructure in the public sector, while contracting out the operating network on a line-by-line basis, as in Stockholm.[56]

In broader terms, urban rail use and provision in Britain has grown significantly in recent years following a long period of relative neglect. The London and the South East region continues to dominate, since the scale of the National Rail and Underground networks there is of a different order of magnitude from those in the provinical regions. Even in respect of light rail systems, London has the largest single concentration of ridership on Croydon Tramlink and the DLR. The high capital cost of rail systems suggests that comprehensive networks in cities outside London, as distinct from selected corridors, are less likely to be provided. An exception might be the Manchester Metrolink, should planned extensions come to fruition, although if this were to be realized, other PTEs would no doubt seek similar treatment of their areas. It could be argued that a government genuinely committed to the realization of a more sustainable transport system would have the establishment of extensive high quality LRT networks in each of Britain's major cities as one of its key goals. Unless the pace of investment is radically increased, however, this is unlikely to be achieved within the period of Labour's *10-Year Plan*.

NOTES

1   Department of Transport (2002) *National travel survey 1999/2001 update*. Transport Statistics Bulletin (02) 22, DfT, London.
2   Greater Manchester Passenger Transport Executive (2001) *Trends and statistics 2000/2001*. GMPTE, Manchester, 16.
3   Gentleman, H; Mitchell, C; Walmsley, D and Wicks, J (1981) *The Glasgow rail impact study*. Transport and Road Research Laboratory Supplementary Report 650. TRRL.
4   Strategic Rail Authority (2002) *National Rail Trends 2001–02, 4*. SRA, London.
5   Social Exclusion Unit (2002) *Making the connections: transport and social exclusion. Interim findings from the Social Exclusion Unit*. www.cabinet-office.gov.uk/seu/publications (accessed 5 November).
6   Knowles, R (1996) Transport impacts of Greater Manchester's Metrolink light rail system. *Journal of Transport Geography*, 4, 1–14.
7   Knowles, R (2000) *Impacts of new light rail transit systems in Canada and Great Britain*. Paper presented at the Annual Meeting of the Association of American Geographers, Pittsburgh, Pennsylvania.

8 House of Commons (2000) Session 1999–2000, HC 153, 8 June, The Stationery Office, London.

9 Begg, D (2001) Trams are us. *The Guardian*, 20 August.

10 Cervero, R (1998) *The transit metropolis: a global inquiry*. Island Press, Washington, D.C.; Knowles, R (1992) Light Rail Transport. In Whitelegg, J (ed) *Traffic congestion: is there a way out?* Leading Edge Press, Hawes, 107–34.

11 Hall, P and Hass-Klau, C (1985) *Can rail save the city? Rail rapid transit and pedestrianization in British and German cities*. Gower, Aldershot; Turton, B and Knowles, R (1998) Urban transport problems and solutions. In Hoyle, B and Knowles, R (eds) *Modern transport geography*, 2nd edition. John Wiley, Chichester, 135–58.

12 Fullerton, B and Openshaw, S (1985) The Tyneside metro in full operation. In Williams, A (ed) (1985) *Rapid transit systems in the UK: problems and prospects*. Institute of British Geographers Transport Geography Study Group, Birmingham, 27–45; Knowles, R (1985) Rapid transit in Greater Manchester. In Williams, A (1985) *Rapid transit systems in the UK*, 46–75; South Yorkshire Passenger Transport Executive (1978) *Transport development plan*. SYPTE, Sheffield.

13 Church, A (1990) Waterfront regeneration and transport problems in London's Docklands. In Hoyle, B (ed) *Port cities in context: the impact of waterfront regeneration*. Institute of British Geographers Transport Geography Study Group, Southampton.

14 Department of Transport (1989) *Traffic in London*. HMSO, London.

15 Knowles, R (1992) Light Rail Transport, 116.

16 Knowles, R (1996) Transport impacts of Greater Manchester's Metrolink light rail system.

17 Haywood, P (2001) *Experience of South Yorkshire supertram*. Paper presented at Light Rail 2001 Conference, Manchester, November.

18 John, D (1992) Tram revival in jeopardy after cuts. *The Guardian*, 10 June.

19 Department of the Environment, Transport and the Regions (1998) *A new deal for transport: better for everyone*. Cmnd 3950, The Stationery Office, London.

20 Knowles, R (1999) Integrated transport, re-regulation and bus quality partnerships. *Transport Law and Policy*, 3, 27–29.

21 House of Commons (2000) Session 1999–2000, HC 153, June 8.

22 Mayor of London and Transport for London (2002) *Where exactly is London's congestion charging zone?* Mayor of London and TfL, London.

23 House of Commons (2002) Session 2001–2002, HC 558–I, 31 January. The Stationery Office, London.

24 House of Commons (2000)) Session 1999–2000, HC 153,55.

25 Department of the Environment, Transport and the Regions (2000) *Transport 2010*, 59.

26 HM Government (2000) *Response from the government to the Eighth Report of the committee: light rapid transit systems*. www.parliament.the-stationery-office.co.uk/pa/cm199900/cmselect/cmenvtra/872/87204.htm (accessed 30 October, 2002).

27 Greater Manchester Passenger Transport Executive (2000) *Metrolink: transforming our future*. GMPTE, Manchester.

28   Salter, A (2002) Metrolink 'big bang' plan hangs in balance: £200m increase in price threatens transport project. *Manchester Evening News*, 6 August.

29   Department for Transport (2002) *£5.5 billion package of transport improvements*. News release 2002/0354, 10 December.

30   Department of the Environment, Transport and the Regions (2000) *Transport 2010: the 10-year plan*. DETR, London.

31   Commission for Integrated Transport (2002) *The Commission for Integrated Transport's initial assessment on the 10–year transport plan*. www.cfit.gov.uk/research/10year/06.htm (accessed 9 June).

32   Begg, D (2001) Trams are us.

33   London Underground passenger statistics are drawn, except where indicated otherwise, from Department of Transport, Local Government and the Regions (2001) *A bulletin of public transport statistics: Great Britain*, 2001 edition. Transport Statistics Bulletin SB (01) 20, DTLR, London.

34   Fares Fair was a policy pursued by the Greater London Council to significantly reduce the cost of using public transport. It was abandoned following a legal challenge from the London Borough of Bromley, which argued that, as no Underground lines ran through the Borough, its rate payers should not be required to subsidise cheap Underground travel.

35   Evans, A and Crampton, G (1989) Myth, reality and employment in London. *Journal of Transport Economics and Policy*, January, 89–108.

36   Newman, P and Kenworthy, J (1999) *Sustainability and cities: overcoming automobile dependence*. Island Press, Washington, DC.

37   See, for example, comparisons in Meyer, W and Dauby, L (2002) Why is rail transport so attractive? *Public Transport International*, October, 4–7.

38   See Willis, J (1997) *Extending the Jubilee Line: the planning story*. London Transport, London; Wolmar, C (2002) *Down the tube: the battle for London's Underground*. Aurum Press, London.

39   Derived from Department for Transport (2002) *Transport statistics Great Britain*, 2002 Edition, DfT, London, Table 5.15.

40   Transit (2002) *London Underground losses rise six-fold as unit labour cost tops £35,000*. 6 September, 18.

41   Depreciation is the process whereby money is set aside in each year, to provide for replacement of assets when they become life-expired. Under accounting practice in Britain this is generally based on the historic value of assets, and thus does not generate sufficient funds to purchase equivalent assets at current prices.

42   White, P (2000) What profit margins are required for the local bus industry in Britain? *Public Transport International*, 6, 12–17.

43   Department for Transport (2002) *Transport statistics Great Britain. Op cit.*, Table 5.15.

44   London Transport (2001) *Annual Report 2000/01*. London Transport, London, 17.

45   A discount rate is the factor applied to future cash flows, whether positive or negative, to express them as present values, allowing for the interest rate that would be paid on borrowings over the corresponding period.

46 The Industrial Society (2000) *The London Underground public private partnership: an independent review*. Industrial Society, London.

47 *Ibid*, Appendix II, *Contract Theory and the Public Private Partnership Proposals for the London Underground Railway System* (by ELSECo Ltd).

48 See Chapter 5 note 41.

49 Newman, P and Kenworthy, J (1999) *Sustainability and cities*.

50 Kenworthy, J and Laube, F (2002) *The millennium cities database for sustainable transport*. UITP, Brussels.

51 Breheny, M (2001) Densities and sustainable cities: the UK experience. In Echenique, M and Saint, A (eds) *Cities for the new millennium*. Spon, London, 39–52. The editors would like to express their condolences to Mike's family and colleagues following his recent passing.

52 Strategic Rail Authority (2002) *Moving forward: leadership in partnership. Annual report 2001–02*. SRA, London.

53 University of Westminster (2003) *Jubilee Line extension impact study*. Transport for London and Department for Transport, London, in press.

54 Strategic Rail Authority (2002) *Strategic Plan*. SRA, London, 79.

55 Strategic Rail Authority (2002) *Strategic Plan*. SRA, London, 79.

56 White, P and Ball, J (2003) Experience of national railways privatisation and of vertical separation in metro systems. *Trasporti Europei*, in press.

7

A 'Thoroughbred' in the Making? The Bus Industry under Labour

John Preston

This chapter reviews the prospects for local bus transport in Great Britain in the light of *A New Deal for Transport* and its 'daughter' documents, in particular *From Workhorse to Thoroughbred*, which aimed specifically to establish a greater role for bus travel.[1] The impacts of the subsequent Transport Act (2000) and *Transport 2010: The 10-Year Plan* on the bus industry are also reviewed.[2] What follows builds on earlier work by considering the recent past and the immediate future of the British bus industry.[3] The chapter begins by sketching out the legislative background and contemporary policy context. It moves on to consider the current role of bus travel in Britain and the mode's potential contribution to the sustainable transport agenda. An examination of the recent trends in the bus markets in London and in the rest of Great Britain is then undertaken. In this context, the chapter reviews some of Labour's recent initiatives and makes some forecasts concerning their possible outcomes. Its main contention is that the 'Workhorse to Thoroughbred' analogy was largely inappropriate and served to distract from the real need to create a greater role for buses within an overall policy of sustainable and integrated transport. It also suggests that developing such a role will need greater financial support than recent governments have been prepared to give.

Legislative Background

The organizational and regulatory structure of the bus industry inherited by Labour was established during the first two Thatcher Conservative governments of 1979 to 1987. The Transport Act (1980) removed fares regulation and liberalized the bus service licensing system, albeit with little impact, but the key legislation was the Transport Act (1985), which abolished the service licensing system for the local bus industry in all of Great Britain except London.[4] This system had existed since 1930. Such deregulation (or, more precisely, quantity deregulation) meant that, provided they followed some basic registration and licensing procedures, commercial operators were now free to provide as much, or as little, bus service as they wished. The Act also introduced a system of tendering for socially necessary services as it was recognized that not all bus services could be operated profitably, particularly early morning, late evening, Sunday and rural services. By 1999/2000, directly subsidized services accounted for 17 per cent of all local bus services outside London.[5]

There were also provisions in the Act for the privatization of the National Bus Company and the commercialization (and eventual privatization) of municipally owned bus companies, such as Darlington Transport or Greater Manchester Buses. At the time of deregulation, some 96 per cent of local bus journeys were provided by publicly owned firms, but by 10 years later this had reduced to less than 10 per cent, with the public sector providing bus services in just a handful of areas (most notably Edinburgh and Nottingham).[6] Extensive consolidation had also occurred, with five major groups (First, Stagecoach, Arriva, Go-Ahead and National Express) controlling some two thirds of the industry. This has led to some concerns that the industry has become spatially monopolized, even though the Act removed the bus industry's exemptions from competition law.

The case for deregulation had been put forward in the 1984 White Paper, *Buses*, and generated substantial academic debate.[7] The reforms were predicated on four propositions. The first of these was that deregulation would lead to substantial movements towards productive efficiency (the production of bus services at minimum cost) and dynamic efficiency (the appropriate rate of introduction of innovative products and practices) leading to unit cost reductions of 30 per cent. This proposition has been supported by subsequent events, with unit cost reductions of 50 per cent achieved (see below). The second proposition was that the bus market was inherently competitive. Moreover, if there were no actual competition this did not matter because the industry was believed to be contestable, where the threat of *potential* competition would ensure that operators behaved in an economically efficient manner (that is, one in which net benefits to society are maximized). If they did not, they would attract 'hit and run' entry. In fact, actual competition has been minimal and the contestability of the industry has proven limited since there are some modest barriers to entry and sunk costs that, in combination with short reaction times, reduce the threat of potential competition. What is more, much of the competition that has occurred has been 'small group' in nature, with no guarantee of an economically efficient outcome.

The third proposition was that the removal of cross subsidy would lead to higher service levels and lower fares on the majority of routes, leading to improved allocative efficiency (the production of appropriate levels of bus service in terms of both quantity and quality, at appropriate prices). We will see later in the chapter that this does not seem to have happened. Although service levels have increased, so have fares, whilst patronage has declined. It is also apparent that there is a strong 'first best' case for subsidy to the bus industry. One of the subtexts of the deregulation package was to reduce subsidy levels, but there was no consideration of what the optimal levels of financial support might be.[8] Moreover, in a 'second best' world where there are constraints on the amount of subsidy available, it can be easily shown that cross subsidy is consistent with economic efficiency. The fourth proposition was that there would be no major external problems, including the impact of privatization and deregulation on congestion and accident rates. With one or two localized and time-specific exceptions, this appears to have been the case.

In essence, the debate which followed the publication of *Buses* was about two contrasting views of regulation. Up to the 1980s, the conventional view was that regulation (and public ownership) could be implemented in the public interest to rectify market failures. It is a view that has generally been supported by the Labour Party, at least since the times of Herbert Morrison who was one of the architects of the regulatory regime that existed between 1930 and 1986. The alternative view, advocated by the Thatcher administrations, was that interventions by government are likely to lead to regulatory failures that outweigh any market failures, at least for industries that are not natural monopolies. This viewpoint notes how regulation often arises from lobbying by particular interest groups and how regulatory policy can be 'captured' by regulated firms.[9] Removal, or dilution, of market incentives leads to productive, allocative and dynamic inefficiencies that are contrary to the public interest. Adherents to this view are deeply sceptical of Labour's current emphasis on an integrated transport policy, which is seen as an attempt to introduce reregulation and central control by the back door.[10]

For London the situation was different, largely because of politics. The then Conservative central government did not approve of the low fares policies being adopted by the Greater London Council in the early 1980s under the leadership of Ken Livingstone (who is the current Mayor of London). Following only limited success with legal challenges, the London Regional Transport Act (1984) took control of most public transport in the capital away from the Greater London Council and placed it under direct government control – a rare example of nationalization in the Thatcher era. Rather than deregulation, a programme of comprehensive tendering was introduced, along with the privatization of London Buses Limited. This process was undertaken gradually, in contrast to the big bang approach adopted outside London, and was not completed until 1994.

Contemporary Policy Context

Labour's aspiration, as indicated by the title of its policy document, was to transform the bus from a 'Workhorse to a Thoroughbred'. The key measures outlined in that document, namely statutory Quality Partnerships – which built on pre-existing but largely informal agreements – Quality Contracts and improved bus services, are detailed below. Other measures included: the provision of better information, both through telephone enquiries (Traveline) and the web (Transport Direct); more joint ticketing (for example, bus – rail integration is being provided by Journey Solutions, a joint initiative between the Association of Train Operating Companies and the Confederation of Passenger Transport); and a national concessionary fare scheme, to provide for travel at least at half the adult fare for the elderly and disabled. There were also a number of specific legislative proposals to make it easier for local authorities to enhance services, to change procedures for service registration and to introduce higher entry standards and more effective enforcement powers.

Further details were provided by the Transport Act (2000) and by *Transport 2010*, the latter setting targets to increase bus passenger journeys by 10 per cent in England. These targets would be achieved through more reliable bus services, provided by better quality, less polluting buses.[11] Over the first five years of the *Plan*, there are expected to be at least 10 bus-based major public transport schemes such as guided bus, around 2,800 miles of Quality Bus Corridors and 1,500 other bus infrastructure measures. In addition, over the first three years of the *Plan*, there are plans for 120 new or extended park-and-ride schemes and 500 voluntary and community transport projects (examples might include the Crawley Fastlink, the A13 Passenger Transport Corridor in Southend and park-and-ride schemes in Plymouth and Salisbury). Exact details of such schemes will emerge as Local Transport Plans and, in some cases, the recommendations of multi-modal studies are implemented (Chapters 3 and 4). An emerging weakness of this approach appears to be that local authorities have limited experience and expertise in developing such public transport schemes.

The Current Role of Bus Travel

Data from the *National Travel Survey* (1999–2001) reveal the current role of the bus in Great Britain. They indicate that the average person makes 59 bus and coach trips per annum (8 per cent of all trips, excluding walking) and travels 343 miles by bus per annum (5 per cent of all trip miles, excluding walking, although other data sources put this slightly higher at 6 per cent).[12] The vast majority of these trips – 57 – are made by local bus services, with a mean trip length of 4.3 miles, whereas the remaining two are made by coach services, with a mean trip length of 49.5 miles. The limited current role of bus and coach services can be contrasted with that of 50 years ago, when they accounted for over 40 per cent of all travel, albeit when the overall volume of travel by all modes was only 30 per cent of that of today. None the less, this still means that bus use today is only half that of 50 years ago.

As is evident from the above, much of the argument concerning bus services has focused on economic factors, but increasing concern with sustainability has led to greater consideration of social and environmental issues, accompanied by calls for buses to be at the core of transport policy.[13] The most important social role for buses is that they provide for those 25 per cent or so of households without regular access to a car. Members of such households make four times as many bus and coach journeys as members from car owning households. It is also important to note that only around 55 per cent of the total population hold a car driving licence. The main driver in households with cars only makes around 20 per cent of the bus trips of other drivers in car-owning households, and only a little more than 10 per cent of the bus trips made by non drivers in car-owning households.[14] A further social benefit is that buses and coaches have an excellent safety record, with an average of only 0.4 fatalities per billion passenger kilometres between 1992 and 2001, compared with 3.0 for cars.[15]

With respect to the environment, buses have a number of advantages. They have lower carbon dioxide emissions per passenger than petrol cars provided they have reasonable loads (40 per cent lower than cars for minibuses up to 16 seats, 130 per cent lower for midibuses with between 17 and 35 seats and 450 per cent lower for standard buses with 36 or more seats).[16] Buses have similarly low carbon monoxide and volatile organic compound emissions, but they perform less well with respect to nitrogen oxide, sulphur dioxide and, particularly, particulate emissions. Improvements in engine technology have, however, made a contribution here: Euro II engines (introduced in 1996) reduced particulate emissions by almost 60 per cent and the introduction of the Euro IV engine in 2005 is expected to lead to particulate emissions that are only five per cent of those associated with Euro I engines. The rapid deployment of new vehicles, along with improved fuel quality and 'end-of-pipe' solutions such as continuously regenerating particulate traps, offer the prospect of eliminating 'black smoke' problems.

London and the rest of Great Britain compared

Labour inherited a two-track bus system in 1997. Outside London, there was a deregulated system in which there was little scope for government intervention, whilst in London there was a regulated system with substantial scope for government intervention, albeit scope that was subsequently devolved to the new Greater London Authority. As a result of its bus industry reforms, the previous Conservative administration had accidentally set up an experiment that permitted the merits of deregulation and comprehensive tendering to be compared. Hardly surprisingly given the circumstances in which it was established, this was not a controlled experiment and comparisons between London and the rest of the country are fraught with difficulties, not least because of different demographic and economic trends (including different trends in car ownership).[17]

Over the last 20 years there appear to have been similar changes in the bus fares index in London and outside London, with an average annual increase in real terms of 1.5 per cent (Table 7.1). Fares increases have been particularly marked in the 1990s. In London, however, there has been the complication of the Travelcard, a relatively inexpensive single-payment ticket which can be used on all modes across the public transport network, introduced in 1983. An alternative measure of fares is farebox revenue per passenger, although this will be distorted by changing trip lengths (Table 7.2). Such a measure suggests a similar level of fare increase outside London over the last 20 years (1.7 per cent a year) compared to the fare index (1.5 per cent), but a lower increase in London (0.6 per cent a year compared to 1.5 per cent). It is probably a more reliable indicator of actual fare trends in London and hence we can conclude that fare increases in the capital have been substantially less than in the rest of the country.

Table 7.1 Local bus services. Fare indices by area (constant prices)

	1980	% change per annum 1980–1989/90	1989/90	% change per annum 1989/90–1999/00	1999/00	% change per annum 1980–1999/00
London	77.4	0.7	82.7	2.4	105.0	1.5
GB outside London	82.5	0.9	90.0	2.1	110.5	1.5

Source: Department of the Environment, Transport and the Regions (2000) *Bulletin of public transport statistics: Great Britain 2000 edition*. DETR, London.

Table 7.2   Local bus services. Farebox revenue (£) per passenger (1999/00 prices). Receipts include concessionary fare reimbursements

	1980	% change per annum 1980–1989/90	1989/90	% change per annum 1989/90–1999/00	1999/00	% change per annum 1980–1999/00
London	0.44	0.2	0.45	1.1	0.50	0.6
GB outside London	0.49	1.3	0.56	2.1	0.69	1.7

Source: Department of the Environment, Transport and the Regions (2000) *Bulletin of public transport statistics: Great Britain 2000 edition*. DETR, London.

Table 7.3 shows that services, as well as fares, have increased. These grew at similar levels inside and outside London in the 1980s, but at much higher rates in London in the 1990s (up 1.8 per cent per annum compared with 0.4 per cent). With respect to demand, at least as measured by passenger journeys, London has massively outperformed the rest of Great Britain (Table 7.4). Demand in the capital was stable in the 1980s, but grew by 0.9 per cent per annum in the 1990s. Outside London, demand declined annually by 2.6 per cent in both the 1980s and 1990s.

Data showing bus operating costs per vehicle kilometre only go back to 1985/86, but over the last 15 years both London and the rest of the country have seen substantial reductions in bus operating costs of around 4.5 per cent each year (Table 7.5). In London the reductions were greatest in the 1990s, whilst outside London they were greatest in the 1980s. In both London and the rest of Great Britain there has been an approximate halving of operating costs over the last twenty years, but with unit costs in London roughly double those elsewhere in the country. There is evidence that bus operating costs are beginning to rise, largely due to less favourable labour

Table 7.3   Local bus services. Vehicle kilometres (million)

	1980	% change per annum 1980–1989/90	1989/90	% change per annum 1989/90–1999/00	1999/00	% change per annum 1980–1999/00
London	279	0.9	304	1.8	365	1.4
GB outside London	1,984	0.8	2,150	0.4	2,234	0.6

Source: Department of the Environment, Transport and the Regions (2000) *Bulletin of public transport statistics: Great Britain 2000 edition*. DETR, London.

Table 7.4   Local bus services. Passenger journeys (million)

	1980	% change per annum 1980–1989/90	1989/90	% change per annum 1989/90–1999/00	1999/00	% change per annum 1980–1999/00
London	1,181	0.1	1,188	0.9	1,307	0.5
GB outside London	5,043	−2.6	3,866	−2.6	2,972	−2.6

Source: Department of the Environment, Transport and the Regions (2000) *Bulletin of public transport statistics: Great Britain 2000 edition*. DETR, London.

Table 7.5   Local bus services. Operating Costs (pence), including depreciation, per vehicle kilometre, adjusted for inflation (1999/00 prices). Figures are net of fuel duty rebate

	1985/6	% change per annum 1985/6–1989/90	1989/90	% change per annum 1989/90–1999/00	1999/00	% change per annum 1980–1999/00
London	296	−5.1	240	−4.1	158	−4.4
GB outside London[1]	159	−9.6	106	−2.7	81	−4.6

Source: Department of the Environment, Transport and the Regions (2000) *Bulletin of public transport statistics: Great Britain 2000 edition*. DETR, London.
Note: 1   Estimated.

conditions. Indeed, the latest data indicate that between 1999/2000 and 2000/01 bus operating costs (including depreciation) per vehicle kilometre in Great Britain increased in real terms by four per cent.[18] The long period of declining unit costs appears to be over.

There have also been substantial changes in subsidy provision (Tables 7.6 and 7.7). Taken together, public transport support and concessionary fare reimbursements have declined each year by 5.4 per cent in London over the last 20 years, and by 3 per cent in the rest of the country. Individually, in both London and elsewhere, concessionary fare reimbursements have grown, whilst public transport support has reduced dramatically, although this support is now beginning to increase. Table 7.7 also shows that there have been significant increases in fuel duty rebate (FDR) in both London and the rest of the country. In total in 1999/2000, bus services in London received subsidy of £171 million, equivalent to 32 per cent of farebox revenue (excluding concessionary fare repayments) and

Table 7.6 Public transport support and concessionary fare reimbursement for local bus services (£ million, 1999/00 prices)

	1980	% change per annum 1980–1989/90	1989/90	% change per annum 1989/90–1999/00	1999/00	% change per annum 1980–1999/00
London	380	−4.5	239	−6.4	124	−5.4
GB outside London	1,119	−3.2	806	−2.7	613	−3.0

Source: Department of the Environment, Transport and the Regions (2000) *Bulletin of public transport statistics: Great Britain 2000 edition*. DETR, London.

Table 7.7 Fuel duty rebate for local bus services (£ million, 1999/00 prices). Estimate based on vehicle kilometres

	1980	% change per annum 1980–1989/90	1989/90	% change per annum 1989/90–1999/00	1999/00	% change per annum 1980–1999/00
London	23	1.2	26	6.1	47	3.6
GB outside London	165	1.1	185	4.5	286	2.8

Source: Department of the Environment, Transport and the Regions (2000) *Bulletin of public transport statistics: Great Britain 2000 edition*. DETR, London.

27 per cent of gross operating costs (before fuel duty rebate). Bus services in Great Britain outside of London received subsidy of £899 million in 1999/2000. This was equivalent to 52 per cent of farebox revenue and 40 per cent of gross operating costs. In some respects, therefore, the system of tendering in London is more commercial than the deregulated regime beyond.

Finally, Table 7.8 summarizes the financial position of the local bus industry in Great Britain in 1999/2000. To the total industry subsidy of £1,070 million, we need to add £500 million for statutory schools transport support (including special needs education), and 'one-off' urban and rural bus grants. The latter include rural and urban bus challenge, rural bus subsidy grant and rural transport partnerships. In England outside London, these subsidy streams accounted for £100 million of the total support in 2000/01. Although these initiatives have undoubtedly made a difference at the local level where new services have been introduced, they have been

Table 7.8   The financial position of local bus services in Great Britain (£ million, 1999/00)

	Revenue	Subsidy	Cost	Profit
London	540	171	624	87
Rest of GB	1,723	899	2,096	526
GB	2,263	1,070	2,720	613

Source: Department of the Environment, Transport and the Regions (2000) *Bulletin of public transport statistics: Great Britain 2000 edition*. DETR, London.

insufficient to make a dramatic impact nationally. Table 7.8 also indicates that up to 60 per cent of the total subsidy manifests itself in operator profits. This has led to concerns that public subsidy is 'leaking' to shareholders because major bus operators have effective monopolies in many areas. This is an interesting twist on the leakage debates of the 1970s and 1980s that focused on how subsidy to the nationalized bus industry leaked into higher wages and reduced productivity.[19]

Great Britain and the rest of the European Union compared

It is instructive to compare the bus systems in Great Britain with those in the rest of the European Union (EU). The recently completed MARETOPE (Managing and Assessing Regulatory Evolution in local public Transport Operations in Europe) project, undertaken for the European Commission, sheds some light on this. The project compared data for bus systems between 1990 and 1999 for up to 31 European cities. Three British cities (London, Leeds and Oxford) were included in the sample. It was found that bus systems in Britain have relatively high cost recovery rates – the proportion of operating costs covered by revenue – of around 85 per cent. Analysis suggests that one of the dominant reasons for this is lower operating costs: earlier work, confirmed by MARETOPE, suggests unit costs were up to 50 per cent lower than elsewhere in the EU.[20] It is also the case, however, that fares are relatively high in Britain – up to double the EU average – and this results in relatively low load factors. Perhaps unexpectedly, it was found that in urban Britain bus use per capita is around the EU average. This is in contrast with results from aggregate statistics which indicate that bus and coach use per capita in the UK is the lowest of all 15 EU member states (Table 7.9). Possible explanations might include very low usage of regular scheduled bus services in small towns and rural areas in the UK or higher usage of nonregular or nonscheduled bus and coach services in the rest of the EU.

Table 7.9   Bus and coach use per capita in the EU, 1998 (thousand passenger kilometres per annum)

B	DK	D	EL	E	F	IRL	I	L	NL	A	P	FIN	S	UK	EU
1.1	2.1	0.8	2.0	1.1	1.0	1.5	1.5	0.9	0.9	1.5	1.4	1.5	1.0	0.7	1.1

Source: Eurostat (2000) *EU Transport in Figures: Statistical Pocket Book*. Office for Official Publications of the European Communities, Luxembourg.
Key: B = Belgium, DK = Denmark, D = Germany, EL = Greece, E = Spain, F = France, IRL = Ireland, I = Italy, L = Luxembourg, NL = Netherlands, A = Austria, P = Portugal, FIN = Finland, S = Sweden, UK = United Kingdom, EU = European Union.

Overall, the European comparisons suggest that the British bus system performs well in terms of cost recovery, but less so with regard to affordability and load factors.[21] This might be seen as a triumph, because very few public transport systems anywhere in the developed world have a cost recovery rate greater than that achieved by the bus system in Great Britain. An alternative viewpoint is that this commercial triumph is an economic disaster that has resulted in less service and higher fares than is optimal. Simulation work suggests that the ideal urban bus system – that is, one which delivers the greatest benefits to the economy as a whole – might only have a cost recovery of between 25 and 50 per cent.[22]

Labour's Current Initiatives

Although overall bus demand has declined in Britain, there have been some success stories outside London. Deregulation has worked well, for example, in Exeter and Oxford, with demand more than doubling in the former and approximately doubling in the latter.[23] Case studies of Quality Partnerships in Brighton, Cheltenham, Edinburgh, Leeds, Leicester, Nottingham and the West Midlands have found patronage growth of as much as 60 per cent, but normally much lower (and there is an important debate concerning how much of this demand is abstracted from other bus services).[24] There have also been some important rural initiatives such as the 'Wiltshire Wigglybus' and the 'Cumbria Plusbus', although these schemes represent a tiny fraction of overall bus use in Great Britain.

The Quality Partnerships that have been studied in most detail predate the Transport Act (2000). They were often informal agreements that appear to have resulted from good working (and personal) relationships between operators and authorities. The Competition Act (1988) seems to have reduced the number of these agreements because of concerns that they might be found to be anticompetitive and result in punitive fines. There was also the worry that Quality Partnerships might be undermined by low fare, low quality operators. Some entry of this kind has occurred – in Leeds and Birmingham, for example – but it has not been long lived. Modelling work suggests that low frequency entry by a low quality operator may be feasible on medium passenger density routes in some circumstances, but that it would be to the detriment of the incumbent operator and, more importantly, wider society.[25]

The Transport Act (2000) attempted to rectify some of these problems. Legally binding Quality Partnerships were given a statutory basis to cover facilities along the route (but not information) and vehicle standards (but not frequency or timing of services). Where operators do not meet the required quality standards they may be excluded from the infrastructure provided under the partnership agreement. The Act also waters down the requirements of the Competition Act (1998) in relation to Quality Partnerships and tendered services, whilst a block exemption for certain joint ticketing schemes came into effect on 1 March 2001. The legislation thus attempts to get over the competition policy and 'free rider' concerns surrounding Quality Partnerships.

Although it is early days, the reforms do not seem to have stimulated an avalanche of proposals. Possible reasons involve the bureaucratic barriers that might exist and, in some parts of the country, a greater need on the part of management to deal with the implications of a much tighter labour market. As the late Michael Beesley observed, entrepreneurial scarcity has often been a problem in the bus industry. Moreover, in a climate of rising tender prices, good relationships between operators and authorities (which are vital for a successful partnership) may become strained.

It is also difficult to envisage how Quality Partnerships can provide lower fares. We have already seen how, compared to the European Union, bus fares in Britain are high and subsidies low. Economic modelling both for Great Britain as a whole and for the Metropolitan areas seems to suggest that fares reductions are desirable.[26] Recent modelling work by Stephen Glaister has indicated that 10 per cent cuts in London and the six English metropolitan conurbations would have net economic benefits. Other recent modelling work, involving the University of Oxford, suggests that fare cuts of 20 per cent could be beneficial for seven specific bus routes, and a study at the national level shows that fare reductions of 30 per cent in London and 40 per cent elsewhere could also be economically justified.

One alternative to Quality Partnerships is Quality Contracts, which would offer an exclusive contract to operate a particular route or set of routes or a network, and would be let by tender. Fares could be specified as part of the contract, and London might be thought of as the prototype. Quality Contracts, however, are at the discretion of ministers and will only be approved if the government agrees that they are the best way of pursuing a local bus strategy. Even if a contract is authorized, there must be a 21 month (in England and Wales) or a 6 month (in Scotland) delay before implementation to allow operators to adjust. It is likely that such contracts would be resisted by the industry, however, given their potential to undermine profitability.[27] A more binding constraint to Quality Contracts might be the lack of skilled planners to design them because of the erosion of professional transport expertise in local authorities (Chapter 3). The design of Quality Contracts might be contracted out to the private sector in the same way as the bus services themselves, although this is not a policy initiative that has so far attracted the attention of Labour ministers.

Another important alternative is performance-based contracts, prototypes of which exist in New Zealand and Norway. They are also being considered for Australia.[28] This might involve awarding an area concession to an incumbent operator, based on a benchmarked subsidy per vehicle kilometre for a minimum service level network that meets Public Service Obligation requirements. There would also be a patronage incentive to encourage ridership based on the social benefits of mode switching. Poor performance (possibly measured by customer satisfaction surveys) could be punished by withdrawing a concession and putting it out to competitive tender. This might provide an alternative to the London system as a basis of Quality Contracts.

One important further issue is the level of subsidy received by local bus services. As already noted, subsidy in 1999/2000 was in excess of £1 billion. The industry is not as commercial as it likes to portray itself. Subsidy constitutes around one third of industry revenue, and this is before statutory schools transport is taken into account. An important question, therefore, is: are we getting good value from this subsidy? The government has recently issued a consultation paper on bus subsidies.[29] In particular it reviews options for the Bus Service Operators' Grant (BSOG, previously the FDR), which refunds around 80 per cent of the duty paid on fuel. The BSOG currently amounts to around £300 million per annum in England. A number of options are considered including: FDR retained but with variable rates for different services (for example, there could be a higher rate for services operated in the evening, on Sundays and in rural areas); FDR retained but with variable rates for cleaner fuels or vehicles (there is already a 100 per cent rebate for gas-powered buses and this approach could be extended); FDR replaced by a per passenger subsidy; FDR replaced by a per passenger kilometre subsidy; transfer funds from FDR to local authority tendering; transfer funds from FDR to rural and challenge funding; abolish challenge funding; and allow the discretionary extension of concessionary fares.

Recent research work for the Commission for Integrated Transport (CfIT) suggests that a move from FDR to a subsidy per passenger would be beneficial, although additional measures to protect rural services that currently have high levels of support per passenger would be needed.[30] There might also be benefits from extending half-fare concessions to the young and low-income groups and in developing specialized services such as American-style yellow school buses. The research also found that free fares for the elderly were generally bad value for money, as were many tendered services and rural transport initiatives. In contrast, support for Quality Bus Corridors and park and ride schemes seemed to represent good value, at least in reasonably sized urban centres.

From Workhorse to Thoroughbred!

We have seen above that between 1989/90 and 1999/2000, bus usage in Great Britain declined by 15 per cent overall, although this was split between a 23 per cent fall outside London and a 10 per cent rise within. If the trends of the past 10 years were to continue for the next decade, then bus demand would further reduce by 13 per cent. This extrapolation is rather crude, as it does not take into account slowing growth in car ownership and use as saturation levels are approached. A more sophisticated estimate, derived from a multi-modal demand model developed by OXERA in conjunction with the University of Oxford, suggests that the decline in bus usage over the next 10 years might be around 5 per cent under a 'do nothing' scenario. Clearly this still falls short of ministers' aspiration of 10 per cent growth across England. But matters are complicated by a 40 per cent growth target for London.[31] If this was achieved within the constraint of 10 per cent growth in England as a whole, a 3 per cent reduction in bus journeys outside London – which is certainly not ambitious and probably not intended – is implied. Recognising this, CfIT and others have suggested a 10 per cent growth target for bus patronage outside London. The OXERA multi-modal demand model forecasts that with the full implementation of the *10-Year Plan* across all affected modes bus use will increase by a little over 15 per cent by 2010, but that this growth will be entirely due to patronage trends in London. Outside the capital, usage is predicted to be stagnant.[32] The government has recently combined the previous targets for bus and light rail patronage and now requires 'an increase in use of more than 12 per cent by 2010 compared with 2000 levels.'[33] This has not addressed CfIT's concerns: as noted in Chapter 6, the increase in London's bus passengers alone is still likely to deliver the new growth objective.

The sustainable transport agenda suggests that a significantly increased role for the bus appears both desirable and achievable. Given the government's current policies, however, it is unclear how this will be delivered. We know that over the last 20 years, bus fare increases have been lower, bus service increases higher, cost and subsidy reductions higher and patronage growth higher in London than in the rest of Great Britain. Whilst acknowledging measurement difficulties and problems with disentangling the impact of external factors, these trends suggest that London is doing something right compared to the rest of the country, and they give some indication of what might be achieved with a regime of Quality Contracts outside the capital. Such a regime may be most appropriate for complex networks with a range of public transport modes and where integrated ticketing may be most beneficial. This might include the areas covered by the seven existing Passenger Transport Executives (PTEs) in Britain's largest conurbations outside London, as well as those covered by possible PTEs-in-waiting in other major urban areas (for example, Bristol, the East Midlands, Edinburgh, South Hampshire, South Wales and Teesside).

Under existing budget constraints, a 15 per cent increase in bus use is feasible.[34] If there are no budget constraints, however, there may be a case for doubling or even trebling subsidy levels (and thereby returning them to those achieved in London and the Metropolitan counties in the 1980s).[35] This would lead to subsidy levels more in line with those in the rest of Europe and North America, and modelling suggests that, if this were to happen, bus use may increase by 50 per cent in the short run, with possibly greater effects in the long run.[36] In short, there appears to be a strong case for increasing government support to bus transport. To some extent, of course, this is envisaged by the *10-Year Plan*, but additional financial resources will generally be directed towards new infrastructure as opposed to revenue support for bus services and there is a danger that Quality Partnerships may result in too much quality at too high fares. Recent work suggests that this may be a particular danger in small towns and on some interurban routes.[37] Moreover, at the practical level, there may be difficulties in delivering quality cost-effectively, particularly as the pool of easily implementable schemes may be rapidly exhausted.

If Quality Partnerships do not have the overall impact on the transport market that the government hopes for, then properly specified Quality Contracts may have a better chance of getting the quality/price combination correct, and also of safeguarding against subsidy leaking to shareholders. The procedural barriers that have been put in the way of such contracts therefore seem unfortunate. An alternative (or indeed complementary) approach might be to review the legislation pertaining to joint-ticketing arrangements to encourage the development of student passes (as in the Netherlands), half-fare discount cards (as in Switzerland) or environmental travelcards, as typified by Freiburg, Germany.[38]

Despite government rhetoric, it cannot be claimed that in recent history the bus, with its 6 per cent market share, has been a workhorse of the transport sector. The days when the bus was a significant mode in passenger transportation are some 50 years past, and policy makers need to work towards re-establishing the bus as a workhorse before they can think of it becoming a thoroughbred. By the 1990s, it had become something of a semiretired pit pony in large parts of the country, a reminder of an industry in near terminal decline. In this context, the aspiration to increase bus patronage seems sensible, not least because in aggregate there is plenty of spare capacity (average loads for local services are around 10 passengers). Yet the growth target for bus use in Great Britain seems extremely modest, especially since the mode was carrying double current volumes half a century ago. Moreover, London, with its proto-Quality Contract regime, is more likely to achieve substantial growth than the rest of the country, where the emphasis in the immediate future will be placed on Quality Partnerships that seem limited in what they can deliver. Some experimentation with Quality Contracts outside London, perhaps in one of the PTE areas or one of the devolved nations/regions, ought to be considered as a priority. Such an experiment might also involve pricing innovations, making use of smartcard technology.

It should also be noted that despite the intentions of *A New Deal for Transport* and the Transport Act (2000), the bus has remained the poor relation of the public transport industry. An important opportunity cost of the post-*10-Year Plan* focus on the railways and on light rail transit may be the gains that could be made from radically improving bus services (Chapters 5 and 6). In this context, the *Plan*'s provision for over £60 billion of public and private investment in the railways between 2001/02 and 2010/11, compared to £58.7 billion for local transport and £25.6 billion for London, is worrying. This is particularly the case given that much of the local transport and London expenditure will not be directly targeted at bus services and will include road schemes, light rapid transit schemes and improvements to London Underground. It seems likely that the bus will remain a poor relation.

There are a number of possible explanations for the bus industry having been neglected. The first is institutional. Delivery of policy is difficult in a fragmented and privately owned industry. The second is political. Bus use is not associated with 'middle England' or the floating voter, and ministers are more likely to talk about 'Mondeo man' than the man on the Clapham omnibus (Chapter 1). The contribution that buses can make to social inclusion may be a policy strength but has political costs. The third is economic. Some decision makers view bus users as less economically important (for example, in terms of spend per shopping trip) than other transport users. The fourth is environmental. Old, lightly loaded buses may not be environmentally benign, particularly if the public health risks associated with particulate matter are taken into account.[39] Although the bus can be a key component of a sustainable transport policy, its credentials in this respect have only been proven in a limited number of cities, most usually when combined with some form of car restraint.[40] To date this has normally been associated with parking charges and controls, but could be associated with road user charging (Chapter 4). If the bus is to be a thoroughbred that actually wins races, then there needs to be a level racetrack and a revised handicapping system. The current handicap system which, at least in congested urban areas, punishes the bus (through too high fares) and rewards the car (through an absence of tolling/road user charging) needs to be reformed. By introducing central area road user charges, London, as in many other respects, may lead the way for the British bus industry (Chapter 4).

In conclusion, rather than continuing to explore a new third way in the form of Quality Partnerships or the introduction of some new rules to the deregulation game, Labour might be better advised to pursue Quality Contracts with more vigour. London has shown that such contracts can provide a favourable third way between deregulated and regulated systems.[41] The London system has also been successfully adopted by a number of cities in the EU, most notably Copenhagen, and it is the main model being considered by the European Commission regarding its proposals to revise regulation 1893/91, which concerns public service requirements and the award of public service contracts for local public transport systems. Admittedly, the London model may not be appropriate for the whole country. It will perhaps work best in the larger conurbations where there is some residual planning expertise and where there is the greatest scope for network benefits. Elsewhere, it might be appropriate to introduce a policy of Quality Partnerships, reinvigorated by a revised and enhanced subsidy regime in which support is redirected from the operation of vehicles to the carriage of passengers. Then, and only then, will the bus industry have a chance of being a thoroughbred in the making. Minor, piecemeal changes of the type that typified the first six years of the Labour administration will not be sufficient.

NOTES

1   Department of the Environment, Transport and the Regions (1999) *From workhorse to thoroughbred: a better role for bus travel*. DETR, London; Department of the Environment, Transport and the Regions (1998) *A new deal for transport: better for everyone*. Cmnd 3950, The Stationery Office, London. It should be noted that, for bus policy, the White Paper refers only to England and Wales. A separate White Paper, *Travel Choices for Scotland*, applies north of the border.

2   Transport Act (2000) *Public general Acts – Elizabeth II*, Chapter 38. The Stationery Office, London; Department of the Environment, Transport and the Regions (2000) *Transport 2010: the 10-year plan*. DETR, London.

3   See Mackie, P and Preston, J (1996) *The local bus market: a case study of regulatory change*. Avebury, Aldershot; Mackie, P and Preston, J (2003) Bus regulation – from workhorse to thoroughbred. In Hine, J and Preston, J (eds) *Integrated futures and transport choices*. Ashgate, Aldershot, in press; Preston, J (2001) An overview of public transport reforms in Great Britain and forecasts for the future. *International Journal of Transport Economics*, 28, 23–48.

4   See Savage, I (1985) *The deregulation of bus services*. Gower, Aldershot.

5   Department of the Environment, Transport and the Regions (2000) *A bulletin of public transport statistics: Great Britain 2000 edition*. DETR, London.

6   In these cities, like others, the bus operators are private limited companies. Unusually, though, they are wholly owned by the local municipalities.

7   Department of Transport (1984) *Buses*. Cmnd 9300, HMSO, London; See also Gwilliam, K; Nash, C and Mackie, P (1985) Deregulating the bus industry in Britain: (B) the case against. *Transport Reviews*, 5, 105–32; and Beesley, M and Glaister, S (1985) Deregulating the bus industry: (C) a response. *Transport Reviews*, 5, 133–42.

8   See Glaister, S (1997) Deregulation and privatization: the British experience. In de Rus, G and Nash, C (eds) *Recent developments in transport economics*. Ashgate, Aldershot, 135–97.

9   Regulatory capture is the 'capture' of a regulator by the industry he or she regulates, and has implications for regulatory independence.

10 Hibbs, J (2000) *Transport policy: the myth of integrated planning*. Hobart Paper 140. Institute of Economic Affairs, London.

11 Department of the Environment, Transport and the Regions (2001) *Transport 2010: meeting the local transport challenge*. DETR, London.

12 Department for Transport (2002) *Transport statistics Great Britain: 2002 edition*. The Stationery Office, London.

13 Pickup, L (1992) Bus policy development: deregulation in a wider context. In Roberts, J; Cleary, J; Hamilton, K and Hanna, J (eds) *Travel sickness: the need for a sustainable transport policy in Britain*. Lawrence and Wishart, London, 243–52.

14 Confederation of Passenger Travel (1999) *The passenger transport industry in Great Britain. Facts. 1999–2000*. CPT, London; see also TraC at the University of North London (2000) *Social exclusion and the provision and availability of public transport*. DETR, London.

15 Department for Transport (2002) *Transport statistics Great Britain*.

16 Romilly, P (1999) Substitution of bus for car travel in urban Britain: an economic evaluation of bus and car exhaust emissions and other costs. *Transportation Research D*, 4, 109–25.

17 Matthews, B; Bristow, A and Nash, C (2001) *Competitive tendering and deregulation in the British bus market – a comparison of impacts on costs and demand in London and the British metropolitan areas*. Paper presented to the Seventh International Conference on Competition and Ownership in Land Passenger Transport, Molde, Norway.

18   See Chapter 6.

19   White, P (2001) Local bus industry profitability and the role of longer-distance services. In Grayling, T (ed) *Any more fares? Delivering better bus services*. IPPR, London.

20   See European Commission (1997) *Improved structure and organization for urban transport operations of passengers in Europe (ISOTOPE). Final Report*. Office for Official Publications of the European Communities, Luxembourg; Holvad, T; Huang, B and Preston, J (2002) On the assessment of public transport regulatory reforms in Europe. Paper presented to the European Transport Conference, Homerton College, Cambridge.

21   See also Commission for Integrated Transport (2002) *European best practice in delivering integrated transport*. CfIT, London.

22   See, for example, Nash, C (1988) Integration of public transport: an assessment. In Dodgson, J and Topham, N (eds) *Bus deregulation and privatization*. Avebury, Aldershot; and Jansson, J (1984) *Transport system optimization and pricing*. Wiley, Chichester.

23   Preston, J (2001) An overview of public transport reforms in Great Britain and forecasts for the future.

24   Bristow, A; Shires, J and Mackie, P (2001) *Quality bus partnerships: new evidence on performance*. Paper presented to the World Conference on Transport Research, Seoul, South Korea.

25   Toner, J; Whelan, G; Mackie, P and Preston, J (2001) *Modelling quality bus partnerships*. Paper presented to the Seventh International Conference on Competition and Ownership in Land Passenger Transport, Molde, Norway.

26   See Mackie, P and Preston, J (2003) Bus regulation; and Glaister, S (2001) The economic assessment of local transport subsidies in large cities. In Grayling, T (ed) *Any more fares?* 55–76.

27   Rodger, M (2002) *Transport delivery in Scotland: the key role that buses can play*. Paper delivered to the Transport Delivery in Scotland conference, The Centre for Transport Policy, Edinburgh, 28 August.

28   Hensher, D and Stanley, J (2002) *Performance-based quality contracts in bus service provision*. Mimeo. Institute of Transport Studies, University of Sydney.

29   Department for Transport (2002) *Review of Bus Subsidies. Consultation Paper*. DfT, London.

30   See Commission for Integrated Transport (2002) *Public subsidy for the bus industry*. CfIT, London; and Faber Maunsell in association with National Economic Research Associates (2002) *Bus subsidy simulation: study final report*. Report for Commission for Integrated Transport, November.

31   Greater London Authority (2001) *The Mayor's transport strategy*. GLA, London.

32   OXERA (2002) Reviewing the *10-Year Plan*. *The Utilities Journal*, 28–29 May.

33   Office for National Statistics (2002) *Bus passenger journeys 2001–02*. News Release   2002/0326.   http://www.dft.gov.uk/pns/DisplayPN.cgi?pn_id= 2002_0326 (accessed 13 December 2002).

34   Mackie, P and Preston, J (2003) Bus regulation.

35   See Glaister, S (1987) Allocation of urban public transport subsidy. In Glaister, S (ed) *Transport Subsidy*. Policy Journals, Newbury, 27–39.

36   Potter, S and Enoch, M (2001) Local transport subsidies and affordable fares: international comparisons. In Grayling, T (ed) *Any more fares?* 1–91; Goodwin, P and Dargay, J (2001) Bus fares and subsidies in Britain: is the price right? In Grayling, T (ed) *Any more fares?* 109–22.

37   Commission for Integrated Transport (2002) *Public subsidy for the bus industry.*

38   Fitzroy, F and Smith, I (1998) Public transport demand in Freiburg: why did patronage double in a decade? *Transport Policy,* 5, 163–73.

39   Romilly, P (1999) Substitution of bus for car travel in urban Britain.

40   Enoch, M (1998) *Bus-based best-practice and urban transport emissions,* unpublished PhD thesis, Energy and Environment Research Unit, Faculty of Technology, The Open University.

41   See Preston, J (1999) An overview of public transport reforms in Great Britain and forecasts for the future.

8

Ubiquitous, Everyday Walking and Cycling: The Acid Test of a Sustainable Transport Policy

Rodney Tolley

Walking is one of the first things a child wants to do, and one of the last things an adult wants to give up, yet as a society we have been taking walking for granted.[1]

When I see an adult on a bicycle, I have hope for the human race.[2]

Sustainable transport is commonly defined as 'transport that meets the needs of the present without compromising the ability of future generations to meet their own needs' (Chapter 1).[3] Given that transport is the fastest growing user of nonrenewable petroleum products, and through this consumption is the fastest growing source of greenhouse gases, our current car-based travel patterns are clearly not sustainable in terms of future generations. Garrett Hardin's 'tragedy of the commons' is an appropriate metaphor, where there are personal payoffs of car use for the individual, but collective disbenefits in that 'critical biophysical systems are stressed beyond capacity and many local and global waste sinks have been filled to overflowing'.[4]

Moreover, the transport we use does not meet our current needs, with congestion worsening in cities throughout the world, and safety, environmental quality, social inclusion, community cohesion, and personal and public health all compromized by the growth in car use. In Germany, in one generation, the increase in car use has not led to more trips, more activities or time saved for the average person. Instead the extra speed conferred by the car is consumed in longer trips, so that people now have to travel further to access the same things that were accessed locally a generation ago.[5] Activities are not more accessible, but people are more mobile, and this increase in the consumption of transport has significant impacts on sustainability. The premise of this chapter is that there is a short- and long-term imperative not only for us to switch to more sustainable forms of transport (which includes public transport) but for large proportions of our travel to be moved to the *most* sustainable forms, walking and cycling.

Despite this necessity, the reality is that since the World War II transport policy in the UK has fundamentally been about accommodating increasing flows of motorized traffic. Walking and cycling have not been seen as alternatives and, indeed, such planning that there has been for these modes has been focused on trying to protect them from the worst impacts of cars and lorry traffic. It is true that gradually their merits have been better recognized, but the promotion of walking and cycling still often takes place in a vacuum, without any serious consideration given to the necessary trade-off between trips by different modes. Given that the number of trips per person is stable at about three per day, it follows that an increase in the number of trips by one mode must be at the expense of another. Therefore the sustainability goal should not be, for example, to increase walking as this might be at the expense of cycling, yielding no sustainability gain. Similarly, increasing walking *and* cycling may be achieved by reducing bus trips, again being effectively 'sustainability-neutral', as car traffic would be untouched.

In sum, it makes no sense in sustainability terms to increase walking and cycling unless, at the same time, car traffic is reduced by a concomitant amount. The goal is less car traffic, not more walking and cycling per se. The key is to reduce trip lengths, which reduces environmental damage per trip and increases the chance of the trip being made on foot or by cycle. Whilst longer journeys need to be switched from car to public transport, it is the 70 per cent of all journeys that are under 5 miles that need to be the focus, as these are either cyclable or walkable – yet are increasingly being driven.

This chapter outlines the benefits of walking and cycling and the obstacles to their widespread adoption. It examines the fluctuating fortunes of the modes in UK transport policy in recent years, and discusses prospects for the future in the light of Labour's approach to walking and cycling since 1997.

The Importance of Walking and Cycling

Until very recently it was difficult to appreciate fully the significance of walking in Great Britain. In fact, the first reasonably comprehensive statistical picture of walking only became available in 1998 with the publication of *Walking in Great Britain*, which demonstrated that although it accounts for only 3 per cent of the total distance travelled, walking is still the dominant mode of transport for short journeys.[6] At least 25 per cent of all journeys are still less than four-tenths of a mile and 80 per cent of these are made on foot. Around half of all education journeys, one third of all shopping journeys and a quarter of social/entertainment journeys are walked. The average person walks about 430 miles per year on public roads.[7] Despite its importance, the amount of walking has declined. In the 20 years prior to 1995/97, the number of walk journeys fell by 10 per cent whilst the average distance walked fell by 24 per cent. This trend has been steepest over the last 10 years and has occurred despite – or perhaps because of – the fact that the average person's total travel mileage has increased by 38 per cent over this same period.[8] Since the mid-1980s the proportion of all journeys made on foot in the UK has fallen from 34 per cent to 27 per cent.

It should be stressed here that much walking is not transport as such, in that people need not be attempting to reach a particular destination to be pedestrians (that is, walking as an 'access mode').[9] Walking is also an 'access sub-mode', as it is the main way of getting to public transport services. In fact, as walking is an integral element of all trips, it may be thought of as the 'oil' that helps the system to run more smoothly. Walking can also be a 'recreation/leisure mode' – including walking for the sake of it, playing and walking the dog. And finally, it can be a 'circulation/exchange mode', where people carry out a range of nontransport activities on foot in public spaces, such as window shopping, chatting to neighbours and friends and having a drink at a pavement café. In other words, walking is not just transport, but a vital part of our use of public space.

Cycling has been in decline in Great Britain since the 1970s, both in the number of journeys made and the total distance travelled, so that only about two per cent of all journeys are now made by bicycle. There are marked variations from city to city, however, with usage as high as 18 per cent of all trips in York. In contrast to everyday use, there are steady increases in the number of people cycling for recreation since the mid-1980s: it appears that many Britons now see cycling as a healthy recreation activity rather than as a useful everyday mode of transport.[10]

As with walking, it is clear that there is great potential for increasing the levels of cycling in Britain. Most trips in urban areas are very short. For example, nearly 70 per cent of all journeys are under 5 miles in length; 43 per cent are under 2 miles and 25 per cent under four-tenths of a mile.[11] Even in the case of journeys undertaken by car, 60 per cent are below 5 miles in length. Average trip lengths to education and shopping are 2.9 miles and 4.2 miles respectively. Commuting distances, though rising on average, are frequently walkable or cyclable by the majority of the population. Fifteen million people own a bicycle, yet only 3.6 million use one regularly. This is one of the measures which indicates that there is considerable suppressed demand.

How Sustainable are Walking and Cycling?

The principal sustainability benefits of cycling are listed in Table 8.1. These benefits also apply to walking, often in an enhanced way. Though this framework is more than 20 years old, it remains relevant and, indeed, some of its contents have become more pressing issues in recent times. In particular, it is the significance of the health benefits of walking and cycling that has been increasingly appreciated over the past two decades, whether conceived narrowly in terms of personal benefits or more broadly in terms of public health. To take walking first, it is the most natural form of physical activity and has been described as 'the nearest activity to perfect exercise'.[12] It can be enjoyed by people of all physical conditions and ages, and requires little in the way of skill or specialist equipment. Evidence indicates that regular walking contributes to reductions in heart disease, diabetes, osteo- porosis, colon cancer, obesity and depression. It is now central to health promotion activity in the UK, not least because, being easy to incorporate into daily lives and split into 'bite-sized chunks', it is actually more likely to be taken up by people and maintained.

The personal health benefits of cycling are, arguably, even greater than walking. Cycling is also an aerobic activity which uses major muscle groups, has the potential to raise the heart rate to an extent that benefits cardiovas- cular health and expends significant amounts of energy. The results have been well documented: regular cyclists are fitter and live longer than non- cyclists.[13] To take just one example, the Copenhagen Heart Study found that those who did not cycle to work experienced a 39 per cent higher mortality rate than those who did.[14]

In many ways the major benefits of more people walking and cycling are experienced by the rest of the population. Firstly, more cycling and walking (at the expense of car use) would significantly improve air quality for everyone and reduce greenhouse gas emissions. Up to 24,000 vulnerable people die prematurely each year and similar numbers are admitted to hospital, because of exposure to air pollution from particulates, ozone, and sulphur dioxide, most of which is related to road traffic (Chapter 4).[15]

A second public benefit advantage is reduced risk of road traffic injury. A disproportionate number of victims of crashes are pedestrians and cyclists, accounting for about 20 per cent of those involved in serious accidents in the World Health Organization's European Region.[16] This is frequently interpreted as meaning that these are dangerous modes, which need there- fore to be reduced if we are to have greater safety. Yet even in the currently hostile environment for cycling in Britain, the benefits from regular cycling in terms of life years gained through improved fitness of regular cyclists outweigh the life years lost in cycling crashes – and the ratio may be as high

Table 8.1   Why plan for cyclists?

It is a cheap way of providing mobility

Cost effectiveness is an important criterion for determining the priorities for investment. Money spent on removing the constraints on cycling can significantly increase the availability of transport for a large section of the community. The provision of routes for bicycles gives people 'freeways' for the price of footpaths

It makes efficient use of space

In congested urban areas, space is a valuable resource. Bicycles take up little space when moving; the capacity of a road is increased approximately ten times if bicycles are used instead of cars. Furthermore, bicycle parking does not rank in the same class as car parking; between 10–15 bikes can be parked in one car-parking space

It contributes to energy conservation

Energy conservation is now a priority in most countries[1]

It keeps people fit and healthy

Regular cycling reduces body weight, reduces heart disease, lessens tension, improves sleep and therefore reduces the cost of health services. It has also been shown to increase people's efficiency at work

It is an equitable means of transport

More people can afford the running costs of a bicycle than any other means of transport and the capital cost is less than a week's wages in most developed countries. The bicycle is a simple piece of machinery to understand and maintain, and thus gives people greater control over their lives

It can cut death and injury on the roads

The annual toll of deaths and injuries to cyclists is appalling. Many are killed and injured through no fault of their own. Cheap and well-tried measures can be used to save lives and injuries

It is a quick means of transport

Door-to-door travel times for urban journeys between 2.6 and 4 miles (the vast majority of urban journeys) can be quicker by bicycle than any other means of transport

It is a reliable means of transport

Bicycles are less likely to break down than other means of transport and are unhindered by traffic jams

It provides mobility to practically everyone

Many of the old, the young and people from other minority groups who will never be able to use cars, can ride bicycles

It is a benign means of transport

The bicycle is noiseless, pollution-free and does not significantly encroach on other people's lives

Source: After Hudson, M (1982) *Bicycle planning, policy and practice*. Architectural Press, London.
Note:
1   A cyclist can travel about 2,500 km on the energy equivalent of five litres of petrol.

as $20:1.$[17] Moreover, if the cycling environment were to be made less intimidating (for example, by reducing traffic speeds and constructing cycle paths and lanes) the benefits would be greater still. Partly this is because existing cyclists would be less likely to come into fatal conflict with cars, but it is also because such improvements would help to release the known latent demand for cycling, and thus save even more lives as more people cycle and fewer people drive.

Thus in contrast to received wisdom, the riskiness of cycling is inversely related to the level of cycle use. For example, the three European countries with the most cycling (Sweden, Denmark and the Netherlands) have fatality rates for cyclists per 100 million kilometres (62.5 million miles) ranging from 1.6 to 2.3. In the countries with the least cycling (Britain, Italy and Austria) they range from 6.0 to 11.0.[18] Clearly, as the amount of cycling increases to a critical mass, traffic arrangements have to be made to accommodate it and car drivers have to adapt their behaviour to share the space with cyclists, producing not only more cycling safety, but more safety for motorized modes too. As one analyst puts it, 'proper planning for cycling and walking is a catalyst for road safety'.[19]

A final public health benefit is that more cycling and walking would lead to greater social interaction and improve 'social capital', a concept being linked increasingly to health. Busy streets sever communities and discourage children from playing, and from walking and cycling to school. By contrast, 'liveable streets' filled with people on foot or two wheels encourage social interaction, build social capital and diminish the fear of crime. These broader benefits of walking and cycling should not be underestimated: 'What use is it to have reduced our risk of death from heart disease and extended our longevity through exercise only to be too scared to walk out of our front doors?'[20]

Obstacles to Walking and Cycling

In order to remove barriers to walking it is important to appreciate the many real or perceived deterrents to walking. Amongst the most important physical deterrents are personal safety, speeding traffic, lack of seating and inadequate pavements (made worse by posts, guard rails, traffic signs, grit bins, sandwich boards, commercial waste, wheelie bins and parked cars).[21] The difficulty experienced in crossing roads is a worsening problem:

> We are corralled behind long lengths of guard railing, forced into dark and dangerous subways and made to endure long waits at pedestrian crossings . . . For once all that has to be done to see the difficulties is to step outside the Palace of Westminster . . . Here in the heart of our largest and

richest city [London], by the nation's best known buildings, it is impossible to cross some of the roads.[22]

Beyond these physical issues, three other obstacles to walking need to be stressed. One is the problem of land-use planning and the location of facilities. Over 90 per cent of walking trips are two miles or less, but distances between homes, shops and schools are increasing. For example, the percentage of homes within a six minute walk of shops selling food fell from 68 to 57 per cent between 1989/91 and 1998/2000. Connected to this is a second issue, that of time, with people feeling that their busy, multitasking lifestyles simply do not allow enough time for them to walk.[23] A third problem is the 'invisibility' of walking. Even though walking accounts for such a large proportion of modal share, particularly for short journeys, it is still often overlooked in policy and in planning. Walking is 'hidden', as it is 'so basic to all planning and transport activities, and so undemanding in terms of government finance, that it somehow slips through the net in strategy formulation'.[24] Walking does not attract big budgets and championing pedestrian issues does not normally further planners' careers.

This lack of professional focus is compounded by the failure of the public to bring the problems walking faces to the attention of planners and politicians. At the national level the Ramblers and Pedestrians' Associations have had little influence on government policy. At the local level, because few people see themselves as pedestrians (compared to those who identify with being drivers or cyclists) pedestrian groups are small, sparsely scattered and lack influence. Transport officers in local authorities cannot justify to their committees the expense of providing for walking, as 'they do not have the active and well informed allies on the outside who will help to demonstrate that "the public" actually wants money spent on facilities for pedestrians'.[25] This lack of really powerful lobby or advocacy groups thereby compounds the problem of the invisibility of walking.

The obstacles to everyday cycling are not intrinsic but are related to the environment in which it takes place. It is an open-air mode which relies on human power and is vulnerable to threat by vehicles. Potential cyclists are deterred by physical barriers such as gradients, heavy traffic and excessive speeds. Moreover, the quality of schemes attempting to overcome these problems has been widely criticized: at a series of seminars on promoting cycling in 2001, practitioners repeatedly noted that the levels of cycling were being depressed by the poor quality of UK cycling schemes, including conflict with pedestrians on shared use paths, lack of continuity of routes, dangerous road junction design and poor cycle facilities at destinations.[26] Four of the five standard cycle route design criteria – coherence, directness, attractiveness, safety and comfort – are frequently ignored, with safety often pursued at the expense of the others by, for example, installing barriers, staggered crossings and routes through subways, all of which make routes less coherent and direct, and more uncomfortable and unattractive. These problems have been ascribed to inadequate guidance from central government, to lack of support from senior management and councillors, and to inadequate training of staff.[27]

It is critical to understand, however, that even top-quality infrastructure – the 'hardware' – will not increase levels of cycling without the 'software' of attitude shifting and behaviour change. For example, the low social status accorded to cyclists has been identified as a major dissuasive factor to cycling.[28] Much research since has re-emphasized this crucial point. Whilst cycle facilities can make cycling safer and are popular, they do not in themselves lead to more cycling. Getting people to consider cycling as a feasible alternative to the car is much more important.[29]

The Changing UK Policy Context for Nonmotorized Transport

Under the Thatcher administrations from 1979–90, support for walking and cycling was almost nonexistent. Writing in 1993, John Adams argued that government transport policy was 'clear, coherent and powerful'. It consisted of promoting an explosive growth in the modern means of travel (cars and planes) and phasing out the 'old-fashioned' means of movement, bicycle, bus and train. The bicycle was, he said, 'clearly heading the way of the horse and cart'. What he called the DoT's 'most impressive achievement – the reduction of walking', was largely undocumented because 'the Department's statisticians did not consider walking to be a form of transport and did not collect information about it'.[30]

Throughout this period, British transport policy was dominated by the belief that motorized traffic growth was an inevitable consequence of economic growth and that it should and could be catered for by increasing road capacity (Chapter 4). But the publication of the *National Road Traffic Forecasts* in 1989, with estimates of road traffic increases of up to 142 per cent by the year 2025, was a catalyst for change. Such traffic could not be accommodated on the existing road network or indeed any conceivable future network. As a result, for the first time in Britain, there grew a widespread realization that there is no possibility of increasing road supply at a level which approaches the forecast increases in traffic. This crucial change in attitude in Britain has been referred to as 'new realism', after an influential review of policy in Britain published in 1991.[31] The implication is that demand management must force itself to centre-stage as the essential feature of future transport strategy, and this must mean an enhanced role for alternatives, including walking and cycling.

A number of reports and events in the early 1990s contributed to a very significant change in the attitude of the government (which by now was under the leadership of Thatcher's successor, John Major). These included the British Medical Association's report on cycle safety, the Earth Summit in Rio de Janeiro in June 1992, the Royal Commission on Environmental Pollution's (RCEP's) report on transport and the environment, and the Standing Advisory Committee on Trunk Road Assessment's (SACTRA's) report which acknowledged the evidence – commonly argued by environmental groups for some years – that new roads tend to generate increased traffic (Chapter 4).[32]

By 1996, definite policy support for cycling was becoming clear as the Department of Transport (DoT) launched the *National Cycling Strategy* (*NCS*).[33] It had targets – for doubling the number of trips by cycle on 1996 figures by the end of 2002, and doubling them again by 2012 – and included a model local cycling strategy for local authorities. The *NCS* was also supported, from the mid-1990s, by a range of policies which have had the effect of raising the profile of cycling, such as school and company travel plans, safe routes to school, new casualty reduction targets, home zones, quiet lanes, traffic calming, government reviews of speed and changes in planning guidance. Within this overall policy framework, air quality and health issues continued to push cycling up the agenda, and the mode received a further boost when the National Cycle Network reached 5000 miles in June 2000.[34]

These gains for sustainable transport were not limited to cycling. In 1996 a discussion document from the Walking Steering Group recognized that to be successful, walking must be considered as part of an overall strategy and not just in isolation. The opportunity to do this came in July 1998, with the publication of the new Labour government's White Paper, *A New Deal for Transport: Better for Everyone*.[35] The 'new deal' for walking and cycling was to make these modes more attractive, safer and therefore more viable alternative modes of transport.

The mid-1990s saw a considerable shift in favour of green modes as, for a short time, motoring groups, environmentalists and politicians seemed to agree that things could not continue the way they were. But, as Lynn Sloman put it, 'almost as soon as the consensus formed, it fell apart'.[36] Specifically, the optimism of *A New Deal for Transport* soon gave way to conflict. Less than a year after the White Paper's launch, a series of demonstrations by road hauliers in protest at fuel prices and road tax blocked motorways and the centre of London. Government responses failed to stress the environmental arguments for the fuel tax escalator, so that protesters' accusations that the government was unjustly taxing motorists were not effectively refuted. Unnerved by its portrayal as 'antimotorist' in hostile tabloids, the government abandoned the escalator (Chapter 1). Further protests and media criticism undermined ministers' belief in the deliverability of the vision mapped out in the White Paper (and, as earlier chapters have discussed, this vision was itself a rather compromised version of the transport ideas which had prevailed in Labour's policy statements in 1996 and 1997).[37] Whereas the White Paper had been based on the need to reduce the rate of road traffic growth, it now seemed to the government as if the only way to achieve this would be by increasing motoring costs substantially. As such action was judged to be politically impossible, ministers began to argue that government policy should tackle the adverse *impacts* of traffic, rather than traffic *volumes* themselves.[38] Of course, for walking and cycling this was a crucial shift in direction, because it opened the door to new road building and meant that the 'noncongestion' consequences of traffic – danger, severance, social exclusion, noise and so on – would get worse, not better (Chapter 4). And so, therefore, would conditions for walking and cycling.

The reversal in Labour government policy after 1998 particularly impacted on walking. Following consultation in 1997 the intention had been to publish a National Walking Strategy (NWS) in April 1998. But by then the government was retreating from its White Paper vision and was becoming increasingly nervous about promoting walking, lest it be seen as 'anti-car'. Ironically, this was compounded by vitriolic media criticism of the then Secretary of State for Transport's decision to travel the 250 or so yards from his hotel to the Labour Party conference venue in 1998 by car. By referring to him habitually as 'Two Jags' Prescott, the tabloids fuelled government concern that the publication of a walking strategy would precipitate Pythonesque 'Ministry of Silly Walks' headlines. The promised strategy was repeatedly delayed and then eventually replaced by an advice document *Encouraging Walking: Advice to Local Authorities*, which was published without a launch or other publicity in 2000.[39]

If the intention was to remove the 'embarrassing' issue of walking from the transport debate, it was conspicuously unsuccessful. The failure to produce a strategy contributed to the House of Commons' Environment, Transport and Regional Affairs Committee deciding late in 2000 to examine the government's record on walking. The Committee identified the need to adopt, firstly, planning policies to promote high density, mixed-use, compact cities which keep distances short and, secondly, recommendations for transport strategies which give priority to and promote walking, produce better conditions for pedestrians, restrain traffic and more effectively manage public space.[40] The government's Social Exclusion Unit was also criticized for realising only belatedly that members of poorer households walk more than others, are more likely to be killed by motorists and lack access to many facilities including shops and medical services. It was felt that tackling these problems in deprived areas could utilize urban regeneration funds, a timely reminder of the cross-cutting nature of many walking issues.

The Committee was, however, clear that the primary difficulty with walking was the attitude to it held by politicians with responsibility for transport. Defending the fact that of the thousands employed by the huge Department of the Environment, Transport and the Regions (DETR), only *two* staff were in post to deal with walking – despite walking accounting for nearly one-third of trips – the Minister of State for Transport said in evidence, 'I suspect it is about right . . . because most of us know how to do it . . . I just think that you can therefore take a lot for granted when it comes to walking'. The Committee tartly observed that 'on this basis a large part of the Government machine could be wound up tomorrow'.[41] Moreover, the Committee noted the widespread criticism of the government for producing an advice note rather than a strategy, which had diminished the importance of walking in the eyes of local authorities, other organizations and professionals. In particular, it had been argued that an NWS would have acted as a trigger to the production of local walking strategies in the same way that the *NCS* had spurred publication of local versions.[42]

Persuaded by these views, the Committee recommended the establishment of an NWS which would indicate, a) the criteria against which local strategies would be examined for the purpose of funding, b) a shift of priorities in respect of policies and spending in its overall transport strategy, and c) how different government departments will coordinate policies to facilitate and promote walking. The Committee was of the view that guidance should be issued under the headings of 'changing priorities', 'funding', 'planning', 'conditions for walking', 'quality of design', 'campaigns to promote walking' and 'research'. The establishment of a National Walking Forum was also recommended, which would exchange best practice, advise on government policy, examine Local Transport Plans (LTPs) (Chapter 3), monitor progress and publish a training strategy.

The contrast between the recognition of the importance of walking expressed by officials from the DETR and the complete lack of political interest drew a plaintive observation from the Committee:

> DETR officials know what should be done. However, as things stand we see little likelihood of progress because Government has not the willed the means to do it. As a result the excellent suggestions in its advice to local authorities, Encouraging Walking, are likely to remain pious but unfulfilled aspirations.[43]

Clearly irritated by the relentless criticism contained within the Report, the government first reacted by trying unsuccessfully to block the reappointment of the combative Committee chair, Gwyneth Dunwoody – who was then promptly voted 'Battleaxe of the Year' in a popular poll – but its formal response, published in November 2001, was more measured.[44] Indeed, the document's upbeat tone on walking took many by surprise. Though it rejected targets and the mooted National Walking Forum, it recommended that local authorities should be encouraged to develop pedestrian-friendly environments, remove guard rails and staggered crossings, observe new policy guidance from government and communicate good practice through a new interactive walking web site. Above all, though, was a significant reversal of policy, which would henceforth be one of promoting the mode, driven by a newly written NWS. Walking, always the Cinderella of transport modes, could apparently go to the ball after all.

Walking, Cycling and the Sustainable Transport Agenda

Current evidence supports both pessimistic and optimistic views of non-motorized transport futures in the UK. On a pessimistic note, the failure so far to make any progress towards meeting the targets set in the National Cycling Strategy, could be cited. In fact, cycling levels rose only by 5 per cent from 1998/99 by 2000 and actually fell by four per cent between the first quarters of 2001 and 2002.[45] Faced with the certainty of missing the target of quadrupling cycling by 2012, the government replaced it in May 2000 with one of increasing cycle traffic to 6 per cent of all journeys by 2010. Explanations for this disappointing outcome could be that it is too early in the process to see tangible results, or possibly the planning and funding arrangements under the new LTP system need to be in place before real change can take place. A more critical view is that the government has demonstrated that it lacks a fundamental commitment to cycling, by failing to match the rhetoric in the *NCS* with appropriate funding.

The funding issue is key because the test of the government's priorities is its spending plans. Of course it is very difficult to identify exactly how much is being spent on nonmotorized modes, because road maintenance, lighting installation or speed reduction measures, for example, may benefit walkers and cyclists as well as motorists. The majority of the submissions on this issue to the Environment, Transport and Regional Affairs Committee claimed that walking was not being made a priority for expenditure. When it was suggested to him that the government puts emphasis on large-scale schemes, the then Transport Minister, Lord Macdonald, agreed. He explained that this is not because they were more cost effective, but instead because their 'impact' and 'value for money' are more easily measured through the methodology 'that is available to the Treasury'.[46]

This was an extraordinary statement. It implied that building large schemes with measured positive value is more worthwhile than building small-scale schemes because their value cannot be measured. Of course, this is nonsense. It is simply untrue that small-scale schemes cannot be measured: indeed there are plentiful examples of such schemes producing very high rates of return in their first year, such as the 526 per cent first year rate of return for the pedestrian priority signals installed in Hull in 1997.[47] More accurately, it seems that Treasury methodologies are being used to conceal the fact that, as the Committee Report comments, 'the government is simply not interested in appraising small schemes', an issue which it said 'must be addressed'. The Committee concluded that the government should ensure that in future funds commensurate with the importance of walking as a mode of transport be spent on the measures put forward in the NWS. The source of this funding was clear: 'there should be a corresponding reduction in the sums spent on new national roads'.[48]

The continuing growth in car ownership is also a problem for the non-motorized modes. Household car ownership is inversely correlated with walking levels and its continued rise provides a strong underlying force for reducing walking. This point was supported in a survey of experts' views of the future of walking.[49] Though this was in a European (rather than British) context, the observation was repeatedly made that although by 2010 there would be more money for walking, more infrastructure, more support and more promotion, underlying cultural trends were deeply unsympathetic. As one expert put it, 'the single most important factor is car ownership levels. These are likely to increase with increased prosperity, and so walking is unlikely to increase in spite of measures to provide better facilities'.[50]

This is another fundamental issue, which exposes the inherent contradiction in Labour's policies on transport. In trying to pursue sustainability yet retain the political support of motorists, a message that car *ownership* should increase but car *use* should decrease has become the norm. Continental countries with higher car ownership levels but lower levels of use are frequently cited, conveniently forgetting that these countries have much better public transport and superior conditions for walking and cycling. Without great expenditure on better facilities for pedestrians and cyclists, one can anticipate continuation of existing trends in Britain where the purchase of a car is a precursor to increased travel distances and sharply reduced levels of walking, in particular.

There must also be doubts about the ability of local authorities to deliver (Chapter 3). A study of their cycle strategies has illustrated the considerable gap between model advice from central government and the approach taken by many authorities. Another study, of local authority policy, has reinforced this view, with authorities blaming the lack of national targets for walking, inadequate guidance documents and design manuals, and a failure to publicise such advice.[51]

Lastly, though there are many splendid examples of revitalized, pedestrianized town centres where walking is attractive and popular, these gains are offset by losses elsewhere, especially due to land-use changes which increase travel distances and make trips less likely to be made on foot or by cycle. The loss of bus patronage, which reduces the number of walk trips to and from the bus stops, is another negative influence. Fear of crime in public places is a further component in the gradual withdrawal of people from public spaces, which reduces the chances of informal supervision so that the street becomes a sterile and more feared place. In her classic text, *Life and Death of American Cities*, Jane Jacobs suggested that the key factor in creating 'liveable cities' is to establish a mixed land-use pattern that decreases distances between facilities.[52] This results in an environment with heavy pedestrian activity, a reduced level of interference from motorized traffic, informal supervision by everyone and a sense of ownership of public space. After all, 'if a city is to be "liveable", it has to be "walkable" '.[53]

Despite the promotoring backlash against sustainable transport policies described earlier, reasons for optimistic scenarios for the future of cycling and walking can be found. First is the fact that an NWS is to be released after all. This will produce more guidance and best practice and will increase the profile of and communication about provision for walking. Funding through LTPs will ensure that local walking strategies will appear for many if not all British towns and cities. At the national level, the Strategy will inform policies in many other relevant government departments and agencies apart from transport. These include those on crime and disorder, health and social inclusion, progress in none of which, arguably, can be delivered without walkable public space.

A second reason for optimism is that the 'new realist' policy prescription initially advocated by green groups has become a mainstream one, the logic of which is widely accepted by transport professionals, academics and even politicians, and that no other intellectually coherent policy package has emerged. In a sense, 'the green movement in the UK has demonstrated that it has the answers'.[54] There is thus a growing understanding of, and support for, green policies in government and local authorities. Whilst politicians can be easily blown off course, government departments change more slowly, and this institutional inertia now seems to be working in favour of the green agenda.

It is unlikely that future events would bear out a wholly optimistic or pessimistic scenario and some combination of the two is more likely, with the balance of wins and losses determining how the progress of the modes is judged. It is, however, possible that the experience of the two modes might actually be different from one another in the future. Indeed, in a way, the prospects for cycling seem less positive than those for walking, as a consequence of the government's redefinition of the transport problem as one of the need to reduce traffic to that of reducing congestion. This means that the future is one of more cars, travelling further if not faster, and given that an enduring obstacle to encouraging cycle use has been the speed and volume of traffic, it follows that getting people to take up cycling in the future is likely to be harder, not easier. It is difficult to see levels of cycling being maintained at existing low levels, let alone tripled, in the face of continuing increases in motorized traffic.

The principal way of reconciling more traffic and more cycling would be to construct a complete network of (primarily) off-road cycle facilities in every British town and city. Technically this is feasible, though it would pose major problems in dense centres. In resource terms too it is possible, as such measures designed to encourage modal switching from the car should logically be funded from what is currently spent in providing for cars. But the real barrier is the lack of political will. Though some cities are exceptions – York and to a lesser extent Edinburgh – the great majority of authorities see cycling as a bolt-on extra to existing transport policies and are in no way ready to countenance the notion of restraining car traffic in order to make more space for cyclists.[55]

In contrast to the situation for cycling, there are policy shifts which have the prospect of making the promotion of walking easier and less contentious. In 2001, the Prime Minister, Tony Blair, gave a speech about local quality of life, or 'liveability', which for the first time acknowledged the importance to people of high-quality public spaces, reduced danger from traffic, and a clean, well-managed, safe and secure street environment: 'Towns, cities, regions and countries that provide safe and attractive places for people to live and work will be the winners. For Britain to prosper, we need to make such places the rule, not the exception'.[56] The primary aim of such urban renaissance is to create people-friendly urban areas by enhancing their quality and environment and, in turn, to improve their amenity, viability and vitality. The creation of safe and attractive pedestrian environments in towns and cities is a necessary condition for success and is central to improving them for shoppers, visitors, workers and residents alike. In other words, quite apart from prowalking arguments based on sustainability, the environment or social inclusion, there is a strong business case for improving walking conditions, as Blair was implying. There is indeed emerging evidence that more walking and cycling produce strong economic benefits for towns and cities.[57] The economic regeneration of Birmingham city centre following its spectacular transformation from a 'motor city' into an attractive place to stroll, linger and shop is probably the best of many examples in Britain.

This changed perspective may allow walking to escape from its local authority 'safety ghetto' and become connected to issues of town centre management, or urban regeneration, whilst at national level powerful alliances can be made with, for example, strategies on social inclusion and crime reduction. Indeed in the future, it may be that the way to promote walking will be to stop talking about walking; that is, rather than promote walking as transport to get somewhere, it may be more effective to promote the things you can do when out walking – window-shopping, strolling, people watching, playing and so on. The focus effectively would shift from the activity of walking to the creation of high quality environments in which walking becomes a natural and pleasurable activity. As an editorial in *Local Transport Today* observed:

> Whilst the public and media are unable to connect with the issue of walking *per se*, paradoxically, they are very concerned to see the conditions that are conducive to walking improve . . . The way to winning the public's heart and encouraging people to step out on the footpaths . . . may be . . . through . . . strategies, which hold resonance in their everyday lives and make walking the natural thing to do.[58]

Interestingly, the Pedestrians' Association, the principal lobby group for pedestrians in Britain, has reinforced this change in perspective by undertaking a major rebranding exercise, in which it has changed its name to 'Living Streets'.[59] At a superficial level, this can be interpreted as escaping from the unfortunate connotations of the word 'pedestrian' and from an image which has demonstrably failed to engage a large percentage of the population. More fundamentally, though, it is a recognition that the key goal of many public policies should be creating liveable public spaces rather than encouraging walking as a means of transport. In the architect Jan Gehl's cryptic phrase, 'there is much more to walking than walking'.[60] The Pedestrians' Association's makeover is a manifestation of the rediscovery of the Roman maxim, *via vita est*: streets are life.

In Conclusion

It might be helpful to preface these concluding thoughts by pointing out differences between the two modes in order to shed further light on future prospects. Carmen Hass-Klau has argued that it makes no more sense to talk about 'walking and cycling' than it does to define a single mode called 'high speed trains and shipping' – yet that is what some traffic engineers do when they aggregate the two modes. It leads, she argues, to bad statistics and superficial thinking. Tim Pharoah concurs:

> The problem is that walking and cycling are very different modes that are in many respects incompatible, and deserving of more careful attention than typically has so far been given. Now is the time to cut the umbilical cord that connects them in the minds of planners and decision takers.[61]

Cyclists are much faster than pedestrians, they use a vehicle and they have a stronger sense of identity as users of their mode – but there are very few of them. In contrast, whilst almost everyone walks, pedestrians are less well represented in the political arena. They do, though, have their own comprehensive, connected, dedicated route network of footways, whereas cyclists are usually expected to share road space with fast and intimidating motorized vehicles, something they are increasingly reluctant to do. Yet pedestrians are also reluctant to help the cyclists out by sharing what space they have with cyclists, which is understandable, given that they have very little to start with and it is increasingly being obstructed by clutter and parked vehicles. Clearly there could be value in pedestrians and cyclists campaigning together for the reallocation of carriageway space away from the car and towards sufficient dedicated space for both pedestrians and cyclists, though there is little sign of this happening. At least the supporters of both modes are agreed that if more space is provided for cycling, it should be at the expense of motorized modes, not of pedestrians.

This, in fact, is the nub of the matter. How is this space to be reclaimed if traffic continues to increase? The issue of substitutability deserves raising again. Cycling and walking both have (or will soon have) their own National Strategies. Cycling has targets and walking does too, in the sense that the government intends to stop the decline and then reverse it. But without increasing the number of trips per person per day, increasing walking and cycling can only happen if some other mode(s) declines. Unless car traffic is restrained, more cycling and walking must come from public transport, which would make people healthier and reduce fuel use and pollution, but would probably reduce the viability of the bus industry and render it even less able to combat the depredations of the car. In short, unless we have targets for all modes, having them for some – even the green modes – is not necessarily supportive of sustainability goals.

In the final analysis, the way a society caters for its pedestrians and cyclists is the acid test of how far it has gone along the way of achieving a sustainable transport system. Without any doubt, there are countless successful projects with nonmotorized components such as home zones, travel plans, safe routes to school, 'walking buses', car free days, walk to work days and much more. Whilst much of this progress is local, it is driven by national guidance and best practice advice, mainly from the Department for Transport (DfT, successor to the DETR and the short-lived Department of Transport, Local Government and the Regions). Yet it also has to be seen against a background of rapidly rising motorization of society and spaces. There are green gains under Labour, but they are set in a sea of red losses. The same DfT is, simultaneously, accepting and anticipating rising volumes of motorized traffic, and this failure to grasp the nettle of traffic reduction sits extremely uneasily with goals of increasing trips by bike and, to a lesser degree, on foot. Indeed, only if Labour shows much greater leadership in directing efforts to reduce traffic are we likely to see a significant shift from the car towards nonmotorized modes. There is currently no sign that it is prepared to do so.

NOTES

1   London Walking Forum (2000) *Walking: making it happen*. London Walking Forum, London, 3.
2   Attributed to H G Wells, source unknown.
3   See Black, W (1996) Sustainable transportation: a US perspective. *Journal of Transport Geography*, 4, 151–9.
4   Hardin, G (1968) The tragedy of the commons. *Science*, 162, 1243–8; Rees W (2003) Ecological footprints and urban transportation. In Tolley, R (ed.) (2003) *Sustaining sustainable transport: planning for walking and cycling in western cites*. Woodhead Publishing, Cambridge, forthcoming.
5   Brög W (1995) *Strategy for the systematic promotion of the use of bicycles*. Social-data, Munich.
6   Department of the Environment, Transport and the Regions (1998) *Transport statistics report: walking in Great Britain*. The Stationery Office, London.
7   Department of Transport, Local Government and the Regions (2001) *Focus on personal travel*. The Stationery Office, London, 16–17; Institute of Highways and Transportation (2000) *Guidance for providing for journeys on foot*. IHT, London.
8   IHT (2000) *Guidance for providing for journeys on foot*.
9   London Planning Advisory Committee (1997) *Putting London back on its feet: a strategy for walking in central London*. LPAC, London.
10 Department of the Environment, Transport and the Regions (1999) *Cycling in Great Britain, personal travel factsheet*. DETR, London.
11 Department of Transport, Local Government and the Regions (2001) *Focus on personal travel*.
12 Morris, J and Hardman, A (1997) Walking to health. *Sports Medicine*, 23, 306–32.
13 Cavill, N (2003) The potential for non-motorised transport for promoting health. In Tolley, R (ed.) (2003) *Sustaining sustainable transport*; Hillman M (1997) The potential of non-motorised transport for promoting health. In Tolley, R (ed.) (1997) *The greening of urban transport: planning for walking and cycling in western cities*. Second edition. John Wiley, Chichester.
14 Andersen, L (2000) All-cause mortality associated with physical activity during leisure time, work, sports, and cycling to work. *Archives of Internal Medicine*, 160, 1621–8.
15 Department of Health (1998) *Quantification of the effects of air pollution on health in the UK*. DoH, London.
16 World Health Organisation (2002) *Physical activity through transport as part of daily activities*. WHO, Regional Office for Europe.
17 Hillman, M (1993) Cycling and the promotion of health. *Policy Studies*, 14, 49–58.
18 WALCYING (1997) *How to enhance walking and cycling instead of shorter car trips and to make these modes safer*. Report 1, 4, Lund University Sweden and Factum Austria.

19   Wittink, R (2003) Road safety has much to gain by the integration of cycle planning. In Tolley, R (ed.) (2003) *Sustaining sustainable transport.*

20   Cavill, N (2003) The potential for non-motorised transport for promoting health.

21   Goodman, R and Tolley, R (2003) The decline of everyday walking in the UK: explanations and policy implications. In Tolley, R (ed.) (2003) *Sustaining sustainable transport.*

22   House of Commons (2001) Session 2000–2001, HC 167–I, 2 July, The Stationery Office, London, ix.

23   Goodman, R (2001) A traveller in time: understanding deterrents to walking to work. *World Transport Policy and Practice*, 7, 50–4.

24   Metropolitan Transport Research Unit (1996) *Putting London back on its feet, the why, how and who of developing a strategy for walking in London.* LPAC and MTRU, London.

25   Gaffron, P (2000) Walking and cycling: does common neglect equal common interests? *World Transport Policy and Practice*, 6, 8–13.

26   Jones, M (2001) Promoting cycling in the UK – problems experienced by the practitioners. *World Transport Policy and Practice*, 7, 7–12.

27   Gaffron, P (2000) Walking and cycling.

28   Finch, H and Morgan, J (1985) *Attitudes to cycling.* Report RR14, Transport Research Laboratory, Crowthorne.

29   Davies, D; Halliday, M; Mayes, M Pocock, R (1997) *Attitudes to cycling: a qualitative study and conceptual framework.* Report 266, Transport Research Laboratory, Crowthorne.

30   Adams, J (1993) No need for discussion – the policy is now in place! In Stonham, P (ed.) *Local transport today and tomorrow.* Local Transport Today Ltd, London, 73–7.

31   Goodwin, P; Hallett, S; Kenny, P and Stokes, G (1991) *Transport: the new realism.* Transport Studies Unit, University of Oxford.

32   British Medical Association (1992) *Cycling: towards health and safety.* University Press, Oxford; Royal Commission on Environment and Pollution (1994) *Eighteenth report. Transport and the environment.* Cmnd 2674, HMSO, London; Standing Advisory Committee on Trunk Road Assessment (1992) *Assessing the environmental impact of road schemes.* HMSO, London.

33   Department of Transport (1996) *The national cycling strategy.* HMSO, London.

34   Lumsdon, L and Mitchell, J (1999) Walking, transport and health: do we have the right prescription? *Journal of Health Promotion International*, 14, 271–9.

35   Department of Transport (1996) *Developing a strategy for walking.* DoT, London; Department of the Environment, Transport and the Regions (1998) *A new deal for transport: better for everyone.* Cmnd 3950, The Stationery Office, London.

36   Sloman, L (2003) The politics of changing to green modes. In Tolley, R (ed.) (2003) *Sustaining sustainable transport.*

37   See Walton, W and Shaw J (2003) Applying the new appraisal approach to transport policy at the local level in the UK. *Journal of Transport Geography*, 11, 1–12.

38 Department of the Environment, Transport and the Regions (2000) *Transport 2010: the 10-year plan*. The Stationery Office, London.

39 Department of the Environment, Transport and the Regions (2000) *Encouraging walking: advice to local authorities*. DETR, London.

40 House of Commons (2001) Session 2000–2001, HC 167–I, 2 July.

41 House of Commons (2001) Session 2000–2001, HC 167–I, 2 July, xxxiii.

42 Gaffron, P (2000), Walking and cycling.

43 House of Commons (2001) Session 2000–2001, HC 167–I, 2 July, xxxiii.

44 HM Government (2001) *The government's response to the Environment, Transport and Regional Affairs Committee's report on Walking in Towns and Cities*. www.local-transport.dft.gov.uk/etrac/walking/index.htm (accessed 28 October 2002).

45 *Local Transport Today* (2002) 15 August, 11.

46 House of Commons (2001) Session 2000–2001, HC 167–I, 2 July.

47 Institute of Highway Incorporated Engineers (2001) *Safe roads for all: a guide to road danger reduction*. Road Danger Reduction Forum, York.

48 House of Commons (2001) Session 2000–2001, HC 167–I, 2 July, xliii.

49 Tolley, R; Bickerstaff, and Lumsdon, L (2001) What do experts have to say about the social and cultural influences on walking futures? *Proceedings*, Australia: Walking the 21st Century conference, 20–2 February, Perth, Western Australia.

50 Tolley, R et al. (2001) What do experts have to say about the social and cultural influences on walking futures? 188.

51 Lumsdon, L and Tolley, R (2001) The national cycle strategy in the UK: to what extent have local authorities adopted its model strategy approach? *Journal of Transport Geography*, 9, 293–301. Gaffron, P (2003) Successful implementation of local authority cycling policies in Great Britain: what does it depend on and how can it be improved? In Tolley, R (ed.) (2003) *Sustaining sustainable transport*.

52 Jacobs, J (1961) *Life and death of American cities*. Pelican, London.

53 Pharoah, T (1992) *Less traffic, better towns*. Friends of the Earth, London, 10.

54 Sloman, L (2003) The politics of changing to green modes.

55 Lumsdon, L and Tolley, R (2001) The national cycle strategy in the UK.

56 Blair, T (2001) Address to Groundwork seminar, Croydon, 24 April.

57 Buis, J (2003) The economic significance of cycling. In Tolley, R (ed.) (2003) *Sustaining sustainable transport*; Napier, I (2003) The walking economy. In Tolley, R (ed.) (2003) *Sustaining sustainable transport*.

58 *Local Transport Today* (2001) 12 July, 2.

59 Pedestrians' Association (2001) *A manifesto for living streets*. Pedestrians' Association, London.

60 Gehl, J (2000) *There is much more to walking than walking*. Keynote address to the first Walk21 International Walking Conference, London.

61 Hass-Klau, C (2001) Walking and the relationship to public transport. *Proceedings*, Australia: Walking the 21st Century conference, 20–2 February, Perth; Pharoah, T (2003) Walking and cycling, what to promote where. In Tolley, R (ed.) (2003) *Sustaining sustainable transport*.

9

Air Transport Policy: Reconciling Growth and Sustainability?

Brian Graham

Air transport is often described – albeit without much substantiation – as being one of the drivers of the increasingly globalized economy. For example, with reference to London, James Simmie and James Sennett state that air transport is one of the key factors promoting the relationship between international trade and this particular innovative city.[1] Again, in an analysis of the ways in which aviation contributes to the United Kingdom's economy, Oxford Economic Forecasting (OEF) highlighted factors such as the industry's roles in facilitating world trade, human networking, increased productivity and improved profitability for other economic sectors.[2] It is estimated that in excess of 0.5 million people in the UK are employed in jobs that are related in some way to air transport and that the industry as a whole is responsible for around 1.4 per cent of total UK GDP. OEF calculates that any curtailment of air transport's growth would strongly disadvantage the UK economy. Assuming a 4 per cent annual growth in air passenger traffic to 2015, a reduction of 25 million passengers per year would mean that by then GDP would be around nearly £4 billion a year lower (in 1998 prices).

Nevertheless, whilst OEF was able to identify a linkage between transport infrastructure and the performance of the rest of the economy, the methodology employed failed to isolate an effect on productivity from aviation separate from that caused by the transport infrastructure as a whole. Thus it might be inferred that whilst air transport plays an important part in national and regional economic growth and development, that role is not necessarily clearly defined nor is the direction of cause-effect relationships readily identified. Further, these economic benefits are achieved at a considerable and as yet unquantified cost. Whilst its case is refuted by OEF, a report for the Institute for Public Policy Research (IPPR) argues that the links between economic growth and aviation have been overstated and are also highly dependent on location.[3] In addition, although air transport creates environmental and social costs through atmospheric emissions and noise, these costs are not internalized by fuel taxation or other fiscal measures. Rather, recent air transport policy, both within the UK and internationally, has been driven by a concern to expedite the growth of civil aviation through the implementation of a competitive global marketplace (albeit one that is widely compromised by the protection of national interests on the part of the most powerful countries).

Thus the emphasis placed on sustainability in the 1998 White Paper, *A New Deal for Transport: Better for Everyone*, signalled a possible change of direction for UK air transport.[4] This was defined as a concept that balanced high and stable levels of economic growth and employment and social inclusion with prudent use of natural resources and effective protection of the environment. In aggregate terms, however, it is readily apparent that the growth trends for air transport in the UK cannot be accommodated within existing infrastructure. Predicted passenger numbers at UK airports in 2015 range from 250 million to 375 million, compared to 181 million in 2001. Even assuming a midpoint forecast of around 310 million, the UK would require the equivalent of 2.25 'Heathrows' by 2015 and another two by 2030. The airport capacity crisis is geographically concentrated in the South East whilst elsewhere many UK airports are actually underused in relation to their available capacity.[5]

A New Deal for Transport refers to the effective use of regional airports in the UK, both as one means of diverting traffic from congested south east England and because of their role in regional development. Regional airports and air services impact on regional economies in three principal ways.[6] First, they function directly as employers: one employee per 1,000 passenger throughput is the 'normal' ratio. Secondly, airports function as catalysts for other on-site economic activities; these can be aviation related but may be businesses attracted, for example, by the accessibility of an airport. Finally, an airport also acts as a regional economic multiplier, although it is exceptionally difficult to measure the degree of embeddedness of an airport in its local economy and to assess the supply linkages and chains to the local economy and employment. It seems logical, however, to argue that smaller regional airports are much less capable of stimulating additional jobs in their regions.

The most detailed assessment of the aggregate impact of aviation on UK regional economies demonstrates that this is a very variable relationship, which is at its most pronounced in the greater South East. In addition to direct employment, it is estimated that aviation creates 1.31 indirect jobs for each direct job at all airports other than Heathrow and Manchester. Further, it is also responsible for induced employment, which is created by expenditure from those employed directly and indirectly by the sector.

Combining all three categories, aviation accounts for 2.9 per cent of total employment in the greater South East, but considerably less elsewhere. Its impact on regional GDP is also very variable, ranging from 0.2 per cent in Yorks and Humber to 3.2 per cent in Greater London. This suggests that all regions do not have the same requirements of air transport and that general conclusions need to be qualified by local circumstances.

In terms of the aggregate relationship between aviation and regional economies, OEF argues that: a) replicating its conclusions about the national economy, the sectors most likely to contribute to regional economic growth are typically those most dependent on aviation, b) good air transport links encourage inward investment, and c) theoretically, improvements in transport infrastructure boost productivity growth across firms that can use it. Whilst these conclusions stem from an econometric model, OEF's empirical analysis of the contribution of aviation to the various regions is weak, drawing together only brief regional profiles and evidence that is, frankly, anecdotal.[7]

This developmental function of regional airports can clearly be at odds with the environmental dimensions to sustainability. For example, attracting firms involved in distribution and logistics is likely to contribute to the congestion externalities already experienced at least by the larger regional airports. More broadly, the hegemony of market forces in aviation is clearly at the expense of the environmental dimensions to sustainability, whilst the economic polices being applied to the air transport industry do not encourage individual constraints in the use of environmental resources on the part of the suppliers of air transport, the airlines. It is also the case that airports, although much more environmentally aware than airlines (if only because it is good business practice to be so), are still working for themselves and not as part of a sustainable system characterized by a need to mitigate the undesirable environmental and social consequences of economic activity.[8] Arguably – the specific issue of surface access apart – both *A New Deal for Transport* and the subsequent government studies culminating in an Aviation White Paper in 2003 emphasize the developmental function of the UK's airports at the expense of the role of air transport in an environmentally integrated transport system. In particular, the government's aviation review has almost completely elided consideration of the National Rail network, even though its deficiencies constitute one of the principal drivers of the growth in domestic UK air transport demand (Chapter 5).

Thus the UK's aviation industry is characterized by rapid but geographically variable growth and is driven by the self-interest of its various stakeholders, particularly on the supply side. There are no overriding national network imperatives and airlines are free to enter and exit markets at will, irrespective of their impact on regional economies. Airlines, but also airports, operate in line with what they perceive as being in their own best interests, a trait most readily demonstrated by the explosive growth of the low-cost or 'no-frills' sector ('low-cost' referring to the operating model which is predicated on minimising costs rather than the fares charged which may often be as expensive as those of more traditional airlines). Meanwhile, rapid traffic growth is overwhelming the ability of technological advances to offset air transport's environmental externalities, principally emissions and noise.

The central aim of this chapter lies in exploring something of this complex of interconnections and tensions that defines the relationships between the growth of air transport in the United Kingdom, its role in regional economic development and the contradictions this presents for the sustainability agenda. A brief discussion of aviation and sustainability precedes an analysis of the role of air transport in UK regional economic growth and development. The discussion then considers the policy levers available to the Labour government in addressing these contradictions, concentrating on the conflicts and tensions between airport planning and the attainment of other broader social, political and economic priorities in a sustainability agenda.

Aviation and Sustainability

In terms of general principles, the UK shares in the global condition that the relationships between aviation and the environmental dimensions to sustainability are largely negative. It has been argued that sustainability as applied to transport has three basic conditions: that the rates of use of renewable resources do not exceed their rates of generation; that the rates of use of non-renewable resources do not exceed the rate at which sustainable renewable substitutes are developed; and that the rates of pollution emission do not exceed the assimilative capacity of the environment (Chapter 1).[9] Air transport, as currently practised, fails outright to satisfy the first two conditions and probably also the third. In the longer term (perhaps 2050+), global air transport is not sustainable on any basis because there is, as yet, no feasible substitute for kerosene, hydrogen-based fuels being the only apparent possibility.

Public opposition to aviation, which is very much determined by proximity to airports and flight paths, tends to focus primarily on noise rather than emissions. Whereas modern aircraft are quieter than their predecessors, it is the volume of traffic – both airside and landside – that compounds public exposure to noise, particularly for residents of the hinterlands of major airports. In general terms, internationally negotiated and implemented noise controls have largely realized the potential returns from aircraft engine noise reduction, and future gains are most likely to come from advances in airframe technology. The International Civil Aviation Organisation (ICAO) is seeking to introduce globally binding 'Chapter 4' noise regulations by 2006. These will be defined as the current 'Chapter 3' rules minus 10 decibels. This is not as large a reduction as it sounds because the 10 decibels is cumulative, being measured at three points, none of which can exceed Chapter 3 limits. In actuality, these regulations will make little difference in the UK or Europe where reasonably young aircraft fleets mostly already comply with the proposed Chapter 4 rules. Noise – or at least its perception – continues, however, to dominate the relationship between an individual airport and its local community, leading to a plethora of often-stringent local operating regulations constraining aircraft operations. One important example is the Heathrow 'QC2' measurement of take-off noise at maximum weight. Even large aircraft, which are Chapter 4 compliant at all weights, cannot make this limit without reduced payloads.

It is becoming more widely recognized, however, that the most serious sustainability impacts of air transport stem from atmospheric pollution at both global and local scales. These include the effects of contrails in the upper atmosphere, carbon dioxide pollution and nitric oxide/nitrogen dioxide (collectively oxides of nitrogen) emissions, all of which contribute to decreases in ozone and to global warming. It is estimated that aircraft emissions have increased oxides of nitrogen 'at cruise altitudes in northern mid-latitudes by approximately 20%'.[10] Because of aviation's growth and the lack of alternatives to fossil fuels, the sector's current 3 per cent contribution to global warming may increase to between 10 and 20 per cent of the total by 2005.[11] In addition, ground-level emissions at airports, both from aircraft and surface vehicles, are increasing, a trend exacerbated by the development of airports as major activity economic centres and intermodal transportation hubs. In general terms, although again technology has been successful in reducing atmospheric and ground emissions per aircraft and per vehicle, the technological returns are diminishing and being offset by aviation's growth.

In addition to its concern with environmental impact, the definition of sustainability also extends to long-term economic development, social needs and equity. Such targets, especially when applied to more peripheral or disadvantaged regions, demand accessibility to core regions as measured by time and cost, whilst the access of isolated areas to wider networks is a basic social equity objective. Firms also require accessibility to factors of production and markets. Yet the infrastructure created to enhance accessibility also encourages mobility, which is essentially a behavioural attribute, and moreover one easily manipulated by price. Arguably it is the provision and ready availability of cheap mobility, best exemplified here by the growth of low-cost airlines, which provides the basic challenge to the environmental dimension of sustainability. This issue is rendered more complex, however, by the democratic nature of low-cost air transport, which allows more people to fly more often. As John Spellar, Minister of State for Transport, remarked in 2002, 'people who call for demand to be managed are saying that workers should be priced off planes'.[12]

Thus even at this most cursory level of analysis, it is apparent that a complex mesh of tensions and contradictions is produced by the relationships between air transport and sustainability. To summarize, policies and strategies that might curb or diminish the environmental externalities of air transport are likely to be swamped by those promoting its development. Although air transport is almost entirely a derived demand, one result of liberalization and privatization is that its provision is now dominated by the interests of private companies, particularly the airlines but also the airports. The latter are more constrained in that they have to interact much more directly through the planning processes with other public concerns and demands than do airlines. Even given this qualification, the marketing strategies of airlines and airports may impact directly and negatively on the attainment of sustainability objectives. These tactics strive to achieve precisely the opposite effect to the curbs on movement inevitably intrinsic to sustainability. They are aimed instead at enhancing air transport demand and increasing volumes of traffic. Airlines and airports as businesses have no rational alternative but to cater to existing markets in ways that generate most profit, whilst fostering future growth. In that conundrum lies the principal dilemma compromising the entire idea of reconciling aviation growth and sustainability, both in the UK and globally.

Air Transport and Regional Economic Development in the UK

Turning to the relationships between the socioeconomic dimension to sustainability and air transport in the UK, at least three sets of processes can be identified. These are: the broad trends in the economic geography of the country; the imbalance in the demand for – and provision of – airport capacity, combined with the inadequacies of existing planning processes; and the previous failure to develop a strategic national policy either for airports or the air transport network, an issue only now being addressed in advance of the 2003 White Paper.

The economic geography of the UK

As is the case in most other Western economies, the UK has experienced a sharp decline in manufacturing employment and a shift in jobs towards the service sector. Business services, financial services and transport/communications account for 26 per cent of all UK employment and the growth trend in these sectors is likely to continue. The 1999 OEF report found that sectors most likely to promote future growth make relatively heavy use of aviation.[13] Therefore the nature of employment in particular regions is a key input into their relationship with aviation and it can be argued that regions capable of specialising in aviation-intensive sectors should perform relatively well in terms of long-term economic growth potential. This is not, however, to imply a direct cause-effect relationship between aviation and regional economic growth.

The UK economy remains dominated by London and the South East, not least because London is the capital city and a global financial centre.[14] By any reckoning, it is a world city and Heathrow remains the world's leading international airport. The UK economy is also now focused on Europe, which accounts for about 60 per cent of all exports; the South East is the gateway to Europe. Whilst the UK has always been conceptualized traditionally in terms of a North–South divide, and although the South as a whole does better than the Midlands and North, parts of it, such as Cornwall, are significantly disadvantaged. Additionally, however, there is also an East – West split. The East has comparatively better rail and road infrastructure linking free-standing cities, whereas the West has declining conurbations around Birmingham and Manchester and is connected to the South East by the clogged M6 and the troubled West Coast Main Line (WCML).

Meanwhile, peripheral regions of the UK, which may be remote or separated by water from the core, are heavily dependent on air transport. Peripherality and remoteness are not, however, objective terms. In this instance, a peripheral region in air transport terms can be defined as one sufficiently distant from the South East to make domestic air travel preferable in journey time and cost to rail or road substitution. Given the state of the UK's surface transport infrastructure, this approximately means north of Manchester and Newcastle. Obviously, its separation by water makes Northern Ireland even more peripheral, air transport being the most effective means of travel to Great Britain and certainly to the South East. The same constraint applies to the Isle of Man and Channel Islands (even if they are Crown Dependent Territories, their air services function as domestic routes). But none of these are remote in the sense in which that term applies to the Inner and Outer Hebrides, northern mainland Scotland, Orkney and Shetland, all of which (Inverness apart) lack direct air access to London.

The imbalance in the demand for and provision of airport capacity

This regionally uneven economic geography means that there is also a spatial variation within the UK in the aggregate demands for air transport capacity and in the requirements placed upon the air transport system. Although their share of national air traffic has declined, Heathrow (61 million) and Gatwick (31 million) handled just over 50 per cent of all the UK's 181 million terminal passengers in 2001.[15] Heathrow is essentially full if measured in terms of available runway capacity at peak periods. Approval for a new fifth terminal (T5), which is intended to allow the airport to accommodate 80–5 million passengers on its existing two runways, was announced in November 2001. Even then T5 will not be in operation until 2008. The numbers, however, which depend on an increase in average aircraft size, do not add up, particularly as British Airways (BA), which holds almost 40 per cent of slots at Heathrow, has been engaged in a strategy of down-sizing aircraft.

There is little alternative capacity in the South East. Gatwick is already the world's busiest single runway airport. Meanwhile traffic at Luton and Stansted has been growing dramatically because of the rise of the low-cost carriers. Passenger traffic at Luton (6.5 million passengers in 2001) has increased by 171 per cent since 1996, the corresponding figure for Stansted (13.6 million passengers in 2001) being even higher at 184 per cent. In total, the five London area airports – Heathrow, Gatwick, Luton, Stansted and City – accounted for no less than 62.6 per cent of all UK terminal passengers in 2001, compared to 65.8 per cent in 1989. The South East contributes over 60 per cent of all origin/destination international passengers. Although the dominance of the region has declined more rapidly in recent years (it handled 64.1 per cent of all passengers in 1998) because of the growth of low-cost carriers at airports such as Belfast International, Bristol, Liverpool and Prestwick, this relatively slight reduction still reflects the long-term and apparently inexorable imbalance in the demand for, and supply of, UK airport capacity.

In marked contrast to this spatial concentration of demand, Manchester, the largest airport outside of the South East, handled only 10.5 per cent of all UK terminal passengers in 2001, compared to 10.2 per cent in 1989. The percentage of passengers using the next top ten regional airports increased from 19.1 per cent in 1989 to 23 per cent in 2001, most of this growth occurring since 1998 because of the expansion in the networks of the low-cost carriers. Although 54 airports in the United Kingdom (and three in the Channel Islands) supported scheduled air services in 2001 (Figure 9.1), in total, the five London area airports, plus Manchester and the next top ten regional airports accounted for 96.1 per cent of all terminal passengers in 2001, compared to 95.1 per cent in 1989. Many of the remaining 38 airports exist only for essential social air service provision in the remoter peripheries of the UK, most notably the Scottish Highlands and Islands.

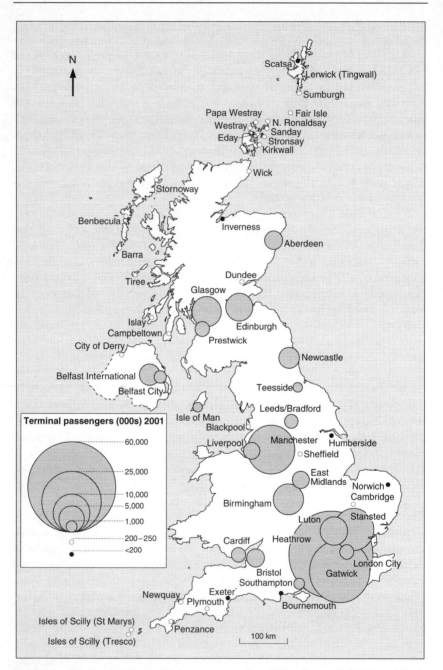

Figure 9.1 UK airports with scheduled passenger services, 2001. Source: Civil Aviation Authority (2002) *UK Airports.* CAA, London.

This largely static picture, relieved only by the traffic growth engendered at some regional airports by low-cost carriers, reflects the inertia built into the UK's air transport system by the geography of demand. The shortage of capacity in the South East cannot be solved by distribution rules or other policies diverting traffic to regional airports, even though there is a tendency for passengers in other regions to increasingly use their own airports for scheduled air travel. *A New Deal for Transport* sought to encourage this trend through proposals for greater liberalization; in essence, the UK has offered open access to all its bilateral partners on international routes to regional airports, provided that UK airlines can operate on the same routes. In practice, however, this is unlikely to have much effect as, Manchester, Birmingham and Glasgow apart, virtually all scheduled international routes at regional airports serve destinations within the European Economic Area (EEA) and are thus already open to any EEA airline following the full implementation of European Union (EU) liberalization in 1997.

Strategic national policy for airports

Until the late 1990s, successive governments failed to elaborate a national strategy for air transport in the UK. The ideologically driven *laissez-faire* Airports White Paper of 1985 and the ensuing Airports Act (1986) were largely concerned with addressing the airport capacity shortage in South East England and establishing the conditions for the privatization and commercialization of the country's airports. Although the 1986 Act excluded airports in Scotland and Northern Ireland, the control of seven airports – Heathrow, Gatwick, Stansted, Southampton, Aberdeen, Glasgow and Edinburgh – was subsequently transferred to BAA plc. More recently, a number of regional airports have been fully or partly privatized. Some airports, most notably Manchester, remain under local authority ownership, not least because they are regarded as key assets in regional development initiatives.[16] Such companies, however, were disadvantaged in seeking capital for expansion because of public sector borrowing requirements. The need for a level playing-field with privately owned airports was recognized in *A New Deal for Transport*, which observed that the Labour government had already announced proposals 'to free soundly-financed local authority airports from public sector borrowing controls'.[17] Exemptions for Leeds-Bradford, Manchester, Newcastle and Norwich airports came into effect in April 1999.

The mixture of ownership within the UK regional airport sector suggests that it is less privatization than commercialization which is changing the nature of the industry. One executive described his airport as 'a runway with a shopping mall beside it'.[18] Rentals from retailing outlets, combined with those from other concessionaires such as restaurants, bars, hotels and car-parking generate far more profit than airside revenues, although the demise of intra-EU duty-free sales in 1999 has had an adverse impact on income. Further, some of the deals being made with low-cost carriers in order to attract passengers actually produce little or no operating revenue for certain airports. Thus it is imperative to the revenues of operators that airports are also planned as catalysts for other forms of on-site development, including business parks and various types of logistics. Such commercialization can clearly accentuate the negative environmental externalities of airports, not least by increasing landside traffic and congestion. Again, each airport is operated as a single company without regard for the wider national interest, priorities and efficiency of the air transport network. In essence, the airport industry is characterized, if less publicly seen as so being, by many of the same problems and inconsistencies that define Great Britain's fragmented and much criticized railway industry (Chapter 5).

Following the publication of *A New Deal for Transport* and driven by the capacity crisis in the South East, an unprecedented series of studies of UK air transport was commissioned in advance of the 2003 White Paper. The government's interim consultation paper on air transport policy, *The Future of Aviation* was published in December 2000.[19] It identifies the key issues and questions that have to be answered in developing a long-term policy for UK aviation. The subsequent White Paper will include the government's proposed policy on the problem of scarce airport capacity and the location of additional infrastructure. It seemed clear from A *New Deal for Transport* that 'predict and provide' had become a dead letter in airport planning, and that there would be some strategy for demand management, even if this took the negative form of not building new runways. Subsequently, the then-Department of Environment, Transport and the Regions (DETR, now the Department for Transport, or DfT) also instituted a series of regional studies, one primary purpose of which is investigate the potential for a diversion of air traffic from the South East to other regions. These were published in July 2002 and they too form part of the consultation process for the White Paper.

The studies set out a range of possible options for the development of airports and air services in the UK over the next 30 years and the policy mechanisms that could be used to deliver them. There are four national growth scenarios: a) RASCO (Regional Air Studies Co-ordination) Reference Case (RRC), which is essentially a predict and provide formula in which both regional airports and those in the South East would expand to accommodate the growth in traffic; b) UK-Wide Constrained (UKC), in which airport development is severely constrained throughout the UK, these limits being reinforced by the use of measures to limit the environmental impact of aviation as much as possible. If this was implemented forecasts indicate that all UK airports would be effectively full by 2020; c) South East Constrained (SEC), in which the development of South East airport capacity is significantly constrained to about 150 million passengers per annum, but no constraints are put on the ability of UK regional airports to grow in line with demand; and d) Facilitating Growth, in which the South East would be generally able to accommodate all of its demand, which, therefore, would not have to be diverted to regional airports. In addition, there are two local spatial scenarios, the specific terms of which vary from region to region. Their generic titles are: a) 'Fly Local', which encourages regional airports to meet as much demand locally as possible; and b) 'Concentrated growth', in which regional growth is focused on one airport so that this can develop a critical mass of travel.

The South East and East of England study (SERAS) identifies the mechanisms required to deliver the RRC and Facilitating Growth scenarios. These include: a third short runway at Heathrow to serve domestic and European routes, which would help increase the airport's overall capacity to 116 million by 2030; as many as three new runways at Stansted, increasing its present 13.6 million passengers on one runway to over 120 million by 2030; a new four-runway airport at Cliffe in north Kent as an alternative to the development of Stansted; and the transformation of the disused military airport at Alconbury, near Huntingdon, into a low-cost airline base.[20] Gatwick was originally excluded from SERAS because of the agreement between its owner, BAA plc, and West Sussex County Council under which the operator undertook not to build a second runway at the airport before 2019. This exclusion, however, was challenged in a judicial review, announced in November 2002, which required the DfT to carry out further projections including a second runway at Gatwick.

The scale of the proposals required to accommodate the upper limit of the government's forecasts has led environmentalists such as Jonathan Porritt to claim that there is no sense of the limits to growth in the consultation documents, a conclusion supported – inadvertently or not – by various statements from ministers.[21] Ignoring the point that the UK is currently ranked nineteenth in the world in terms of *per capita* income (a primary determinant of the demand for air transport), the Secretary of State for Transport, Alistair Darling, stated: 'Doing nothing is not an option . . . We have built the fourth largest economy in the world on our ability to trade. Air travel is crucial to our expanding economy'.[22] Darling also reiterated the argument that the UK requires at least one world-class airport hub to compete with its European rivals, Paris Charles de Gaulle (CDG) and Amsterdam Schiphol, which already have much better runway provision than Heathrow, and Frankfurt Main where a third runway has been approved.

Ultimately, however, the 2003 White Paper, albeit further delayed by the Gatwick decision, should define a national policy for airport capacity. A positive benefit is that this may help alleviate the existing situation dominated by short-termism, expediency and an absence of strategic planning. This reflects the conclusion of *A New Deal for Transport* that whilst UK airports are operating increasingly in a free market, 'each airport cannot be viewed in isolation from other airports. Airports both compete with each other and complement each other to some extent'.[23] The final decision will clearly rest between the RRC/Facilitating Growth and SEC scenarios and probably with the Concentrated Growth spatial scenario, which makes more sense in terms of airline network economics. Whilst the SEC scenario would comply better with a sustainability agenda, many commentators would agree with Jonathan Porritt that there has been a move away from the ethos of the 1998 White Paper and even evidence that the Labour government is reverting to a predict-and-provide mentality. Certainly, it seems that the upper limits of the RRC and Facilitating Growth scenarios are being accepted as the *actual* forecasts for future growth, and the UK aviation industry sees this consultation process and the subsequent White Paper as being its best ever chance of getting the infrastructure that it craves. Nevertheless, huge environmental barriers would have to be overcome if a version of the RRC/Facilitating Growth scenarios was to prevail. One example of the potential conflicts involved is provided by the plans for Heathrow, the government's own estimate being that as many as 15,000 homes housing 35,000 people would have to be demolished by 2015 around the airport if a third runway was built, unless significant technological advances were made in the reduction of oxides of nitrogen emissions to comply with the limits set in EU regulations.[24] Even if unconstrained growth was the preferred strategy, it seems very unlikely that, because of local environmental issues, all regional airports would be allowed to indulge in uncontrolled growth.

In addition to SERAS, DfT's Regional Air Services studies also cover Scotland, Northern Ireland, the north of England, the Midlands, Wales and the South West.[25] These investigate the impacts of the national policy scenarios on the regions and identify the key national and local issues specific to each. Although most projected growth can be accommodated through modification of existing infrastructure, the scenarios could include: a new airport at Bristol, the replacement of Birmingham airport by a new facility between Coventry and Rugby, an additional terminal at Manchester and an extra runway at either Edinburgh or Glasgow. Whilst the various regional studies reveal much about the relationships between air transport and the regions, and environmentally related issues such as surface access, the particular issue of air transport and regional economic development remains tangential if only because the primary questions centre on the capacity of the air transport system.

The air transport network

One important factor compromising the regional studies is the lack of input from the suppliers of air transport, the airlines, businesses that pay scant regard to the concept of a national air transport network and are responsible for continuous churn in the provision of air service as companies enter and exit regional markets at will. Three recent examples from Northern Ireland illustrate this point:

- immediately following 11 September 2001, BA announced the abandonment of its Heathrow – Belfast International route (six daily), the slots being reallocated to intercontinental services simultaneously transferred from Gatwick to Heathrow;
- bmi British Midland, which had announced the switch of its Heathrow service from Belfast International to Belfast City from 28 October 2001, largely because low-cost operators were being given preferential deals, then partially rescinded this decision and maintained a split operation through the summer of 2002 with a seven-daily frequency from Belfast City plus five daily from Belfast International. The latter has been reduced to three daily for the 2002–3 winter, the probability being that the service will be dropped entirely for the summer schedules in 2003, leaving Belfast with seven or eight daily frequencies to Heathrow compared to as many as 15 in 2001;
- in August 2002, easyJet, beset by self-inflicted scheduling problems created by a new crew rostering system that did not work properly, gave the public precisely four days notice of the immediate termination of those services between Belfast International and Scotland flown by go (with which it merged in 2002). Prospective passengers, all of whom had already paid for tickets, were accommodated on other easyJet flights but at the *airline's* convenience.

These illustrations of the self-centred nature of airline business strategies also have resonances for the current key issue in the provision of UK regional air services, the decline of the full-service scheduled domestic carriers as the low-cost airlines – such as the merged easyJet/go and Ryanair/buzz, and bmibaby and MyTravel Lite – expand their activities (and portfolio of hideous brand names). There is considerable variation in the low-cost impact on domestic routes. In some markets, such as the London area to Scotland or the North West (particularly Liverpool), the carriers expanded the traffic base. In other cases, as on the London area to Northern Ireland routes, they have simply captured traffic in mature markets from the traditional carriers. Most low-cost carriers now seem to be evolving into an easyJet model, which is aiming at the business market and major airports, whilst maintaining low costs, compared to a Ryanair variety in which low price is all and the product – mobility – is being sold as one element in disposable consumer income. This model also often involves the use of secondary airports (for example, Prestwick for Glasgow) although that is less true of the company's UK and Ireland services than of its continental network. Irrespective of the model, the low-cost carriers can be seen (and depict themselves) as enfranchising those previously excluded from using air transport because of its cost. Thus it can be argued that they do fulfil a serious social equity role, even though altruism is scarcely a guiding principle in their business plans. Conversely, the low-cost carriers are a major impediment to the attainment of an environmentally sustainable air transport system because they encourage mobility for its own sake. A succinct (and forthright) definition of what might be described as 'mindless mobility' is provided by Bob Geldof: 'If I can get a £7 flight to somewhere within two hundred miles of Venice, well I'll fucking well take it. Seven quid, I don't care where I fucking go'.[26]

In terms of UK regional economic development and accessibility, the key weakness of the low-cost carriers is that they are point-to-point operators, which, given the decline of traditional full-service carriers, may be fine for domestic services, but means that peripheral regions may lose connectivity to European and intercontinental destinations across Heathrow, the UK's premier international airport. This is particularly the case if the traditional operators – such as BA – withdraw from domestic markets in the face of the low-cost competition, which they have done so much to encourage through their own high-cost pricing policies. None of the low-cost carriers offers through ticketing, even for passengers connecting to another of their services. Moreover, for a regional airport an existing low-cost point-to-point service to Amsterdam or Paris may effectively preclude the introduction of what might be a more valuable hub connection service on the same city-pair.

Most UK regions, especially those more distant and/or peripheral, have excellent point-to-point services to the London area's five airports (Heathrow, Gatwick, Stansted, Luton and City). The low-cost carriers (hitherto concentrated at Luton and Stansted) have been especially significant in this market and there is ample evidence that domestic business travel is gravitating towards these airlines, which also cater for a large leisure market. Nevertheless, although transfer passengers may be a minority market – for example, about 72 per cent of Northern Ireland-London Area traffic is point-to-point – Heathrow remains the UK's principal international airport for connecting traffic.

Only nine UK regional airports are connected to Heathrow, however, and there has been a consistent pattern of outright loss of service or diversion to Gatwick (Table 9.1). Around 20 regional air services to the South East have been lost since 1985, the key factor being the capacity constraints at Heathrow which mean that scarce slots can be used more profitably for

Table 9.1   UK and Channel Islands regional airports: connections to London area, September 2002

Regional airport	Heathrow	Gatwick	Luton	Stansted	London City
Aberdeen	★	★	★		
Belfast City	★	★			★
Belfast International	★		★	★	
Dundee					★
Edinburgh	★	★	★	★	★
Glasgow	★	★	★	★	
Guernsey		★		★	
Inverness		★	★		
Isle of Man		★	★		★
Jersey		★			★
Leeds–Bradford	★	★			
Londonderry				★	
Manchester	★	★		★	
Newcastle	★	★		★	
Newquay		★		★	
Prestwick				★	
Plymouth		★			
Teesside	★				
TOTAL	9	13	6	9	5

other markets. British Airways has consistently reallocated its own domestic slots to more profitable international services, switched services to Gatwick and bought slots from other domestic operators (Guernsey lost its Heathrow route when the then Air UK sold its slots to BA). However, although Gatwick retains numerous international flights, BA's failed strategy to make it a second London hub means that it is not an efficient connecting point. To some extent, Manchester and Birmingham offer limited connectivity but the former, in particular, is very high cost compared to London. Peripheral areas like Northern Ireland and Scotland are thus highly dependent on services to Europe's most congested airport – Heathrow – for their international connections, whilst others are forced to rely on Gatwick. Even then, the survival of Northern Ireland's links to the world is entirely dependent on the future business strategy of one airline, bmi British Midland, which is 49.9 per cent owned by the Star Alliance carriers, Lufthansa and SAS. There is no guarantee that some of the airline's Heathrow slots – by far its most valuable assets – will remain dedicated to UK domestic services when Star operates on a global network with vastly different requirements.

This raises the issue as to how regions might protect their remaining air connections to Heathrow, where, given the capacity crisis, airlines may switch slots to other more profitable services. The difficulty is that there are very few policy mechanisms through which regional administrations or even central government can intervene (even if the latter was minded to do so). One prime exception is the European regulation for the provision of Public Service Obligation (PSO) air routes, although this has previously been interpreted as applying only to routes that would not otherwise be flown without subsidy.[27] Whilst the UK, unlike other EU countries, has been very reluctant to use this legislation, it is now investigating the possibility of declaring PSO routes that would not attract subsidies. Previously, the States of Jersey and Guernsey have tried unsuccessfully to invoke this regulation to protect the islands' ill-fated Heathrow services, whilst the Scottish Executive is pursuing this course with relation to the Inverness-London service. One difficulty is that the regulation refers to city-pairs and not to airport-pairs.

In ensuring good air access that could bring wider benefits to regional economies and communities, it might also be the case that regions should be allowed to buy slots at Heathrow, which could be leased to airlines on a transparent financial basis. Regional authorities could also invest in air services although this would have to be compliant with EU rules on competition and subsidies. It proved possible, however, for the South West Regional Development Agency in association with Cornwall County Council and Restormel Borough Council, the joint owners of Newquay airport, to invest substantial capital into the Ryanair Stansted-Newquay service, which began in April 2002. Regions also benefit from access to competitive air services and thus PSO provisions might be extended to cover routes between regional airports and hub airports in continental Europe. As Frankfurt Main is heavily slot-constrained, this essentially means KLM's Amsterdam and Air France's Paris CDG hubs, although a Brussels service is also desirable because it connects regions to the EU locus of power.

Air transport and regional development: a summary

Given the pressures exerted on the UK's air transport system by the belated attempts of national policy makers to address the imbalance in the demand for – and the provision of – airport capacity, what are the repercussions for the socioeconomic connotations of sustainability? *A New Deal for Transport* and the 2002 Regional Consultation Documents argue that the growth of traffic at regional airports can contribute to a sustainable transport policy, largely by contributing to regional regeneration, whilst offering the possibility of relieving congestion in South East England and reducing the need for long surface journeys to those airports, which handle the bulk of international traffic to and from the UK. The scope is limited, however, as only Manchester and, to a significantly lesser extent, Birmingham, offer a credible range of international services although virtually all UK regional airports are served by connections to alternative European hubs at Amsterdam and Paris CDG. Indeed more UK regional airports are connected to Amsterdam than to Heathrow, whilst Paris CDG has the same number of services (Table 9.2). These routes can be seen as essential if they substitute for a Heathrow service, or protect against its withdrawal.

Although it does seem logical to suppose that good air links are beneficial to regional growth, the evidence is ambiguous. It is very difficult to establish the causality between the expansion of an airport and wider economic development, and Anne Graham notes that the quantification of multiplier and catalytic effects is problematic.[28] There are more positive data from North America where Keith Debbage points to the interconnectedness of aviation and urban economic change.[29] It may be the case that there are significant variations in this regard between North America and Europe with its more constrained geography and alternative forms of inter-city transport.

One of the most detailed investigations of the relationships between air transport and regional economic development in the UK is a study carried

Table 9.2   UK and Channel Islands regional airports: connections to European mainland hubs, September 2002

Regional airport	Amsterdam Schiphol	Paris Charles de Gaulle (other Paris)	Frankfurt Main (Hahn)	Brussels National (Charleroi)
Aberdeen	★			
Belfast International	★			
Bristol	★	★	★	★
Cardiff	★	★		★
Edinburgh	★	★	★	★
East Midlands	★	★		★
Glasgow	★			
Guernsey	★			
Humberside	★			
Jersey		★		
Leeds–Bradford	★	★		★
Liverpool	★	(★)		(★)
Manchester	★	★	★	★
Newcastle	★	★		★
Norwich	★			
Prestwick		(★)	(★)	(★)
Southampton	★	★		★
Teesside	★			
TOTAL	16	9 (2)	3 (1)	8 (2)

out for the Northern Ireland Department of Regional Development (DRD).[30] Its conclusions are surprisingly ambivalent. On one hand, it is comparatively easy to establish that air transport is both a significant direct employer and an important provider of freight connections. It is also of vital importance to the in-bound tourism industry, since Northern Ireland is at a particular disadvantage in an all-Ireland context because of its lack of direct low-cost services to mainland European cities. In terms of inward investment, however, air services do not seem to have played a pivotal role in locational decision-making. Both inward investors and indigenous firms are taking full advantage of the recent arrival of low-cost carriers on Northern Ireland routes. Belfast International has the highest exposure to this class of traffic among all UK regional airports, virtually 80 per cent of its total passenger throughput travelling on low-cost airlines. Notwithstanding this development, firms are still concerned by costs. Although the lack of direct flights to destinations outside the UK is not seen as a major problem (an easyJet service to Amsterdam currently constitutes the sum total of Northern Ireland's scheduled air connections to non-UK destinations), London Heathrow is – inevitably – regarded as the principal international hub.

The key issue that emerges has resonance for other UK regions. Low-cost airlines have opened up the Northern Ireland market and more trips are being made per capita for reasons that include attendance at sporting fixtures, cultural events and shopping, as well as the more traditional business trips and visiting friends and relations. If, however, the low-cost carriers were to squeeze out the airlines providing interline facilities, especially at Heathrow but also at Manchester and Birmingham, this would quite drastically limit the range of destinations accessible from Northern Ireland, thereby bringing the argument back to the government's apparent reluctance to engage with the PSO issue.

Policy levers available to the Labour government

The Northern Ireland study confirms, too, that decision making in the provision of services within the UK air transport market increasingly rests in the hands of airlines, involuntarily abetted by airports desperate to gain traffic through the growth of low-cost carriers and anxious not to lose out to potential rivals. This illustrates a basic paradox in the interaction between free market aviation and a sustainability agenda. The environmental dimension to the latter is inevitably downplayed whilst the suppliers of air transport do not necessarily fulfil the socioeconomic objectives of regions. The Northern Ireland example demonstrates that regions require stability in the provision of air services. But aircraft are infinitely moveable assets and the UK domestic market is characterized by continuous churn as companies respond to long- and short-term crises (some of their own making) by constantly changing route networks, schedules, aircraft capacities and frequencies.

It is thus readily apparent that a reconciliation of air transport growth and sustainability objectives in the UK is exceptionally difficult, not least because the environmental and socioeconomic dimensions to sustainability in themselves create fundamental contradictions. *A New Deal for Transport* recognized these problems in its very limited discussion of air transport but failed to isolate any processes through which they might be resolved. The reason is simple enough, in that even the most rudimentary analysis points to an absence of policy levers through which central government and devolved regional administrations might intervene in the relationship between air transport, regional economic development and sustainability. The European Commission (EC) has competence in all matters relating to air transport liberalization, competition and privatization. Consequently, safety regulations excepted, central government can do little about airlines although it does have more influence in airport policy. However, whilst air transport is a reserved issue (that is, retained within the Whitehall remit), land-use planning in the UK is the responsibility of local and regional administrations. Consequently in Scotland, Wales and Northern Ireland, airport planning is, *de facto*, a devolved issue although the regional administrations have no competence in any matter relating to airlines, bar environmental regulations, the monitoring of which is the responsibility of individual airports. Leaving aside international negotiations on air service agreements with other sovereign states, the principal lever available to central government in England is the land-use planning system, which can be used to prevent, constrain or facilitate the development of airports. In addition, central government can intervene through: various financial measures that might encourage or constrain investment in aviation-related infrastructure, the introduction of noise and emission targets at various airports, regulating the availability of landing and take-off slots at airports, the supply of air traffic control and airspace capacity.

The pre-eminence of land-use planning among the policy levers for aviation has several important consequences. Because UK planning in general is devolved to the regional and local scale, every airport, no matter how large, has to interact with its local community. Although national and regional economic development priorities are often of vital importance, the airport-environment interface is at the core of all negotiations with local communities and of planning decisions.[31] *The Future of Aviation* refers to the need to streamline and modernize planning procedures for major projects but argues that statements of national policy have to be established before further major projects can be considered.[32] Obviously Heathrow T5 is the precedent to be avoided at all costs. The planning inquiry took over five years and, even then, the terminal will not open until 2008 at the earliest. The overall process from conception to completion of the terminal will take – at best – an astounding 16 years. Clearly over such extenuated time periods and in an industry as dynamic as aviation, the rationale of the project may change radically, as do the predictions and growth trends underpinning the business model on which it is based. In such circumstances, airports have inevitably evolved their own individual strategies to avoid planning inquiries. One of the most common is the idea of partnerships between stakeholders in air transport and local communities, whose representatives meet in Air Transport Forums. BAA plc, the major UK airport operator, now has a policy of negotiating what are essentially contracts with the communities around its airports as a means of avoiding public inquiries. Such strategies maintain the traditional UK approach of achieving environmental objectives through negotiation and persuasion.[33]

The relationship between air transport and planning includes at least five potential causes of conflict. They are: a) the ubiquity of sustainability issues; b) the problem of reconciling the needs of the air transport industry with continued public participation in the planning process; c) the scale of planning policy-making and the actual and potential conflicts of interest between the priorities of the agencies operating at different scales; d) spatial variations in planning demands, priorities and strategies; and e) the overlap between airport planning and other land uses.

Sustainability issues

As is already apparent, sustainability issues are at the core of all airport planning policy making. The UK planning system is arguably not particularly well structured to achieve any resolution of the inherent contradictions of sustainability. It really only decides on the appropriateness of land use and location and does not effectively deliver sustainability judgements.[34] Yet the economic, social and environmental costs and benefits of aviation to society are at the heart of the debate on airport planning.

Public participation in planning

To some extent, the difficulties in UK aviation planning reflect the wider problem that land-use planning offers the only institutionalized opportunity for environmental and community pressure groups to participate in environmental policy making and contest the interests of the producers. Thus the inquiries become the focus of this conflict, as they constitute the principal arena to address the fundamental contradictions between environmental protection and economic development.[35] Otherwise, it can be argued that the interests of the producers are not readily compromised by planning procedures characterized by democratic deficit. John Mohan points to the growth of an unelected entrepreneurial state in the UK, the basic function of which is to override the interests of local communities in the attainment of national economic growth.[36] Government, whether regional or national, is only one element of a much more extensive range of institutions and actors involved in airport policy making. These stakeholders include agencies of the unelected state (primarily those involved in regional development), various quangos and public private partnerships, pressure groups and even the Ministry of Defence. (The latter, whilst attempting to sell redundant military airfields such as Farnborough into civilian use, is not subject to requirements of planning permission at all.) These processes are driven by central government, one example being the OEF report with which this chapter began.[37] That was funded by the airlines and DETR and, in arguing for the importance of aviation to the UK economy without taking into account the externalities of air transport, illustrates what Mohan sees as a key unresolved tension within the UK between a democratic agenda and one focused substantially on economic development. The uncontrolled growth of low-cost airlines and the conflation of their self-interest with the public good provide another example of this tension. Mohan goes on to argue that the presumption in the UK planning system is essentially in favour of development, the onus of opposition being firmly placed on objectors such as environmental policy groups who remain as 'intermittent insiders'.[38]

The scale of planning and policy making

The difficulties created by the diversity of the governance process are compounded by the question of scale of policy making. Development planning for airports – as all else – is essentially a local function, negotiated in local circumstances. Therefore the primary agencies involved are the local planning authorities. Then, however, there are unelected and unaccountable institutions with broader spatial portfolios, such as the Regional Development Agencies, who see airports as drivers in regional economies and identities. The evolving UK national policy to be revealed in the 2003 White Paper also has to be congruent with the broader remit of the European Commission over many aspects of the aviation industry, and even with globally binding agreements on noise.

Spatial variation in planning within the UK

Planning priorities for airports also vary spatially within the UK. Compare, for example, the extraordinary saga of the Heathrow T5 inquiry with the uncontroversial development of the DHL air freight terminal that has helped make East Midlands the third ranked airport for cargo in the UK after Heathrow and Gatwick. One reason for such variations is that environmental policy in local authorities is often an outcome of the actions of environmentally conscious individuals occupying important positions of influence and power. More broadly, the application of 'deep green' policies may reflect the success of local pressure groups with the ability to dictate local agendas.[39] One example of this local bargaining, which delivers a solution agreeable to both community and airport operator and avoids a planning inquiry, is the 1979 agreement to maintain London Gatwick as a single-runway airport until 2019. As this indicates, the socioeconomic composition of an airport hinterland – in this case wealthy Sussex – is also an important factor in the planning process. Attitudes to airport planning further reflect other spatially variable processes such as uneven development, patterns of new-firm formation, regional labour markets and even the form of airport ownership. It is also probably the case that airport planning is expedited in a region with a strong identity because it can be portrayed as a gateway and even as a symbol of the dynamism of that region. One example is the close relationship between Manchester Airport and the north west of England.

Airport planning and other land uses

Obviously, airport planning can also overlap and conflict with the Planning Policy Guidance (PPG) for other land uses. For example, PPG24 on Planning and Noise does not stop residential developments being built around airports, even within the contour that defines the limit for noise insulation. Again, airports are multi-modal transport interchanges but they are also multifunctional locations. Despite the obvious environmental externalities that can accrue, PPG13 on transport does not preclude airport diversification, including business uses, logistics and distribution, industry and even, somewhat paradoxically, housing development. Airport planning has also to be seen in the context of its wider effects on airport hinterlands and green belts, whilst airfields are essentially brownfield sites, even if they are in open countryside.[40]

Towards the reconciliation of economic growth and sustainability

This complex raft of issues raised by the relationships between air transport and planning (summarized in Figure 9.2) demonstrates that government policy for aviation is compromised by its dependence on a mechanism that

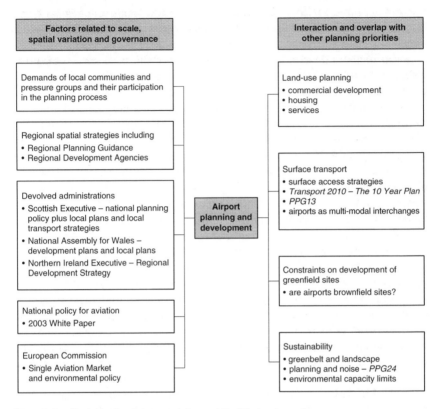

Figure 9.2   The interaction between aviation and the UK planning system.

is both indirect and often subverted by other priorities. Nor is the relationship between air transport and regional economic development defined by self-contained cause-effect processes. Rather it is mediated through this multifaceted set of interconnections with numerous other aspects to the political and social planning structures of the UK and its devolved realms of governance.

Moreover, the absence of dedicated policy levers means that the UK's emerging aviation policy is focused on airports and their role in a national network, whereas the suppliers of air transport – the airlines – operate in an environment of unaccountable self-interest whilst using (often) publicly funded infrastructure. Airlines claim ownership of slots at capacity re-stricted airports, even though these have largely accrued though 'grand-father rights' (that is, previous use). One think-tank, IPPR, has advocated an annual auction of slots to force airlines to use them more efficiently.[41] Such proposals reflect the crucial problem that government lacks any effective means of legislating for airline business activities. Thus there is a conflict of public and private welfare, the provision of airport capacity being the only really significant dimension along which the two interact. Meanwhile, although air transport emerges as an enabling rather than a causative factor in regional development, regions require stability of service from the airline industry but have also very few mechanisms to ensure the protection of their air services.

So is it possible to reconcile aviation and sustainability? *A New Deal for Transport* did seem to signal a possible change in direction but, as with the other transport modes discussed in this book, the environmental and socio-economic dimensions to sustainability seem less prominent in more recent documents. In part, this reflects the tensions within the notion of sustainability itself. On the one hand, the projected growth in UK air transport and, in particular, the role of low-cost carriers in stimulating mobility without necessarily enhancing accessibility, is incompatible with mitigating the effects of the environmental damage created by air transport. On the other, many UK regions, especially those more distant from the South East, would benefit from the construction of additional airport capacity in the London area. For Scotland or Northern Ireland, promotion of the socio-economic dimension to sustainable development would clearly benefit from a third runway at London Heathrow, that being the most promising means of protecting their interlining potential and maintaining access to the world. Good interconnectivity potential at Heathrow should be a *sine qua non* for the regions, even if that means – in any eventual absence of a third runway – the introduction of PSO routes, hitherto anathema to successive 'free market' governments. Again, this is a question of airlines being publicly accountable for their ownership of slots, assets that they have acquired largely without payment.

It is in the South East that the environmental dimension to sustainable development is most likely to come into play. None the less, the air transport industry is lobbying hard and it feels that the government is listening. Even if the latter opted for either the RRC or Facilitating Growth scenarios and the massive political backlash which that would cause, there are few indications in the consultation documents as to how the increased airport capacity would be distributed or which airlines would use it where. BA, for one, is firmly committed to Heathrow. If, conversely, the SEC scenario prevails, it will represent *de facto* demand management by virtue of not supplying sufficient capacity for projected air traffic growth and could thus be presented as a sustainable compromise. One key issue is whether airport capacity should be provided to encourage mobility when that demand could often be satisfied by environmentally more sustainable surface transport.

The current balance of power in the UK air transport business privileges the interests of the airlines in that the Labour government lacks policy instruments – and perhaps the political will – that could ensure that this form of transport does deliver on its potential contribution to economic development. Moreover airlines are not accountable in the way that airports are through the planning process. Whilst that is not a flexible or particularly efficient set of policy levers for air transport, it does have the virtue of providing a democratic arena in which the industry is forced to respond to legitimate public concerns about the impact of air transport on the environment. Thus any reconciliation of growth and sustainability requires airport capacity constraint. Airlines should not be allowed to dictate the agenda because their self-interest is not of prime concern to the overall welfare of society. Air transport is important for regional growth but the interconnections in this process are complex and certainly not amenable to simple cause-effect reductionism. Nevertheless, if there is to be capacity constraint, and thus by default a sustainable agenda for aviation, it is incumbent upon government to use the mechanisms available to ensure that accessibility for regions is not denied through the actions of privately owned airlines acting in their own self-interest. Why should the air transport industry expect to be exempted from state intervention for social and political reasons when it forms part of a broader transport infrastructure that serves need and in which other modes are used for precisely these purposes?

NOTES

1   Simmie, J and Sennett, J (2001) London: international trading metropolis. In Simmie, J (ed.) (2001) *Innovative cities*. Spon Press, London, 191–228.
2   Oxford Economic Forecasting (1999) *The contribution of the aviation industry to the UK economy*. OEF, Oxford.
3   Oxford Economic Forecasting (2001) *Comments on IPPR discussion document 'Sustainable aviation 2030'*. www.oef.com (accessed 29 October 2002); Grayling, T and Bishop, S (2001) *Sustainable aviation 2030*. IPPR, London.
4   Department of the Environment, Transport and the Regions (1998) *A new deal for transport: better for everyone*. Cmnd 3950, The Stationery Office, London.
5   Graham, B and Guyer, C (2000) The role of regional airports and regional air services in the United Kingdom.
6   Graham, B and Guyer, C (2000) The role of regional airports and regional air services in the United Kingdom.
7   Oxford Economic Forecasting (2002) *The economic contribution of aviation to the UK: part 2 – assessment of regional impact*. OEF, Oxford.
8   See Upham, P (2001) A comparison of sustainability theory with UK and European airports policy and practice. *Journal of Environmental Management*, 63, 237–48; Graham, B and Guyer, C (1999) Environmental sustainability, airport capacity and European air transport liberalization: irreconcilable goals. *Journal of Transport Geography*, 7, 165–80.

9   Greene, D and Wegener, M (1997) Sustainable transport. *Journal of Transport Geography*, 5, 177–90.

10   Intergovernmental Panel on Climate Change (1999) *Aviation and the global atmosphere*. Cambridge University Press, Cambridge, 31.

11   Intergovernmental Panel on Climate Change (1999) *Aviation and the global atmosphere*. Cambridge University Press, see Chapter 2.

12   Address to Freedom to Fly Conference (2002), TUC Congress Centre, London, 24 July.

13   Oxford Economic Forecasting (1999) *The contribution of the aviation industry to the UK economy.*

14   Simmie, J (ed.) (2001) *Innovative cities*; Simmie, J; Sennett, J; Wood, P and Hart, D (2002) Innovation in Europe: a tale of networks, knowledge and trade in five cities. *Regional Studies*, 36, 47–64.

15   All figures are taken from the relevant annual reports of the Civil Aviation Authority, *UK Airports*, CAA, London.

16   Humphreys, I (1999) Privatisation and commercialisation: changes in UK airport ownership patterns. *Journal of Transport Geography*, 7, 121–34.

17   Department of the Environment, Transport and the Regions (1998) *A new deal for transport*, 78.

18   Humphreys, I (1999) Privatisation and commercialisation, 128.

19   Department of the Environment, Transport and the Regions (2000) *The future of aviation: the government's consultation document on air transport policy*. DETR, London.

20   Department for Transport (2002) *The future development of air transport in the United Kingdom*. South east and east of England regional consultation document, DfT, London.

21   Address to Freedom to Fly Conference (2002).

22   *The Times* (2002) 24 July.

23   Department of the Environment, Transport and the Regions (1998) *A new deal for transport*, 77.

24   *The Times* (2002) 2 August; *The Guardian* (2002) 21 September.

25   Department for Transport (2002) *The future development of air transport in the United Kingdom*. Regional consultation documents: Scotland, Northern Ireland, North of England, Midlands, Wales and the South West, DfT, London.

26   Cited in Calder, S (2002) *No frills: the truth behind the low-cost revolution in the skies*. Virgin Books, London, 73.

27   European Regulation 2408/92.

28   Caves, R and Gosling, G (1999) *Strategic airport planning*. Pergamon, Oxford; Graham, A (2001) *Managing airports: an international perspective*. Butterworth Heinemann, Oxford.

29   Debbage, K (1999) Air transportation and urban-economic restructuring: competitive advantage in the US Carolinas. *Journal of Air Transport Management*, 5, 211–21.

30   Oxford Economic Forecasting and Airport Strategy and Marketing (2002) *The Northern Ireland air services study*. OEF, Oxford and ASM, Manchester. For Department of Regional Development, Northern Ireland.

31   Graham, B and Guyer, C (1999) Environmental sustainability, airport capacity and European air transport liberalization.

32   Department of the Environment, Transport and the Regions (2000) *The future of aviation.*

33   Carter, N and Lowe, P (1998) Britain: coming to terms with sustainable development. In Hanf, K and Jansen, A-I (eds) *Governance and environment in western Europe.* Addison, Wesley, Longman, Harlow, 17–39.

34   Mohan, J (1999) *A United Kingdom? Economic, social and political geographies.* Arnold, London, 203.

35   Carter, N and Lowe, P (1998) Britain.

36   Mohan, J (1999) *A United Kingdom?*

37   Oxford Economic Forecasting (1999) *The contribution of the aviation industry to the UK economy.*

38   Carter, N and Lowe, P (1998) Britain, 34.

39   Muir, K; Phillips, M and Healey, M (2000) Shades of green in local authority policy-making: a regional case study. *Area,* 32, 369–82.

40   Gallent, N; Howe, J and Bell, P (2000) New uses for England's old airfields. *Area,* 32, 383–94.

41   *The Guardian* (2002) 10 August.

Part III

The Future

10

Towards a Genuinely Sustainable Transport Agenda for the United Kingdom

Phil Goodwin

The concluding chapter in a book like this needs to acknowledge the detailed analysis and evidence in the various chapters, and seek to offer a way forward which balances the evidence for pessimism and the desire for optimism. The reader will by now be completely aware of the main single question, which unites the various analyses of the chapter authors: has the government abandoned, for reasons of political expediency, the integrated and sustainable transport strategy it had promised? It follows: if so, when did the U-turn take place? What were the roles of the 1998 White Paper, *A New Deal for Transport: Better for Everyone,* and *Transport 2010: the 10-Year Plan?*[1] How do we explain the evolution of transport thinking over this period? And what can happen next?

I am not an entirely disinterested observer on these questions, having chaired the panel of advisers that helped to write the 1998 White Paper, then later substantially criticized the *10-Year Plan* as incompatible with many of its objectives. Other authors have seen the White Paper itself as already having abandoned promises made earlier. The government accepted, in December 2002, that the *10-Year Plan* would not deliver even those modest claims made for it, especially in its highest profile promise, to reduce congestion, and was essentially a programme to 'slow down the pace at which things get worse' rather than to make genuine and substantial improvements. The ramifications of that acceptance are profound.

With some important caveats, the other authors in this volume broadly find themselves on the same side of the argument. The book has not tried to be representative of all possible viewpoints in the transport policy debate, but rather those which accept that the concept of sustainability is meaningful and important, that transport is an important part of it, and that solutions depend – among many other factors – on accepting that the scale and character of historical trends in growth of movement have to be managed, or damped or reversed.

Not everybody involved with transport policy accepts this. Probably the greatest area of consensus relates to road traffic in the heart of urban areas, where there is real evidence in many countries of the advantages of traffic reductions, especially in city centres, and where voices for major large-scale urban road building, intended to match and provide for unrestricted traffic growth in cities, are in advanced Western countries mostly confined to a small fringe of disregarded fantasies. At the other extreme, ideas of active intervention to curb the growth of international movement by air carry as yet little popular support, and hardly any accepted practice outside a few medium distance routes for which there are good high-speed rail alternatives. The Royal Commission on Environmental Pollution (RCEP) has raised the fundamental long-term unsustainability of 'predict and provide' as a model for air traffic, but the alternatives, and instruments for achieving them, are not yet remotely as clear as for surface transport.[2] In the middle we see debates on interurban travel (where traffic growth and major road building are still live policy prescriptions, albeit often divisive and contested), uncertainties over whether major long distance public transport investment is a symptom of the problem or an essential part of the solution, and discussions on – for example – policies on ports or goods deliveries which are more often conducted in terms of commercial practice than public policy.

A Problem for New Labour, or a Problem for Government?

Many (though by no means all) of those concerned with preparing the way for Labour's transport policy when in opposition were naturally Labour Party activists, or close to the party in other ways. As a result, they have tended to see the evolution of transport policy in the context of the parallel development of 'New Labour' and its distinctive concerns and political style. It naturally follows that those who find one disturbing tended to relate it to the other, and there is an argument that one might describe, or cartoon, as 'New Labour had abandoned the socialist principles of the Labour Party, and this is the reason why the government has abandoned the related principles of the 1998 Transport White Paper'.

There are some connections that can be made, and the one that resonates most with traditional Labour Party concerns is the decision not to renationalize the rail industry, followed by the complex treatment of Railtrack, then Network Rail. It is also true that the word 'integration' has had a long importance in Labour transport policy, albeit with major shifts in interpretation, from the manner of governance of a nationalized transport sector in the 1940s, to concern about coordinating bus and rail timetables and ticketing in the 1970s, to recognition of the interactions between modes, and with land-use planning, now (Chapter 1).

But to see the current problems as being distinctively Labour would miss the whole point. This is not a problem of any particular political party, but of government. It parallels similar problems discussed in many other countries, led by whatever party. And most important, while the language of the manifesto and *A New Deal for Transport* clearly used the language of the Party, the concepts are simply not to be understood this way. This is essential if we are to understand what has gone wrong, or indeed whether it has, and why, and what can be done about it.

The proof is historical. As Chapter 1 noted, the key turning point in bringing local and global environmental questions effectively into transport discussions was a Conference of the Ministers of Transport of 19 European countries, in November 1989. The conference was chaired by the then Minister of State for Transport, Michael Portillo, and received a series of expert reports on the extent to which transport was an important contributor to environmental pollution. The picture that emerged is now reasonably well known. In summary, the transport sector is one of the major polluters, and the largest sector for which environmental damage was growing, largely due to increases in private car ownership and use. The European Conference of Ministers of Transport adopted (unanimously) a resolution which went considerably further than any previous multinational statement of its form.[3] Britain signed up. The language and substance were quite remarkably close to what was later embodied in the 1998 White Paper, as may be seen in the following extracts:

Governments should review the use of taxes and/or regulations for motor vehicles to ensure their consistency with the goal of reducing fuel consumption and emissions.

A full range of possible measures that can be taken to reduce transport's contribution to the 'greenhouse effect' be set out together with the costs and practical problems of implementing them.

Traffic management [should] be used to further environmental objectives in transport policy, both in relation to demand management and in relation to changing modal split.

it is necessary, in accordance with the Polluter Pays Principle, to introduce systems of supplementary charging for environmental damage caused.

effective and acceptable means of reducing the use of the private car in urban areas need to be applied.

assessments of infrastructure investment proposals should include traffic and environmental evaluations of the alternatives, including . . . extending railway or other public transport infrastructure and that of not building the infrastructure.[4]

The next major landmark was the White Paper *This Common Inheritance*, published in September 1990 by eleven government departments, including Transport, and led by the Department of the Environment.[5] At the early stages of preparation of this White Paper there had been some media attention to comments by the Secretary of State for the Environment, Chris Patten, about the Department of Transport's traffic forecasts, leading to an expectation that the Department of the Environment might distance itself in some way from an 'unacceptable' future of massive increases in car ownership and road provision. In the event, Cecil Parkinson, the reluctant Secretary of State for Transport, had already made it clear that 'the forecasts are not a target . . . it is *not the Government's aim to cater for all forecast demand in all circumstances* . . . there will be cases, for example in city centres, where on economic or environmental grounds – or indeed both – it is neither practical nor desirable to meet the demand by building new roads'.[6]

The White Paper stated:

> It is simply not possible to cater for unrestricted growth of traffic in our city centres, nor would it be right to accept a situation in which traffic congestion found its own level, with inefficient use of road space and increased fuel consumption.

The approach advocated was:

> taking pressure off unsuitable routes and allowing environmental improvements, together with improvements to traffic flow on the strategic road network and improved public transport with greater priority for buses . . . in most cases it does not make economic or environmental sense to increase capacity on roads leading into already congested areas simply to facilitate additional car commuting.

Concerning road pricing, there was no great enthusiasm but also a clearly deliberate refusal to give firm rejection:

> Eventually it may be necessary to consider rationing use of road space by road pricing, but this approach is largely untried and there would be difficulties in ensuring an enforceable and fair system.[7]

The 1989 roads White Paper, *Roads for Prosperity* – which contained, according to ministers, the 'largest road building programme since the Romans' – was progressively marginalized as its schemes were abandoned, or referred to further study (Chapters 1 and 4).[8] In this process, an important political influence was thinking at local government level, especially in those areas which would have to bear the brunt of traffic growth on local roads, accelerated by expanded trunk roads, and which did not have the same possibilities for road construction: what is the point, it was asked, of expanding motorways just to increase congestion in the surrounding areas?

For the rest of the 1990s, a string of scientific and research-based reports had the effect of underpinning and increasing confidence in the tentative policy reorientation described above. One recurrent theme was the identification of specific elements of best practice in other countries which had progressed further than in the UK (pedestrianization of city centres in Germany, public transport investment in cities in many countries, experiments with road pricing in Scandinavia, traffic calming in the Netherlands and so on). Some landmarks in this development – all initiated by, or during, Conservative administrations, but continuing following the change in government in 1997 – were:

- a British Road Federation report which demonstrated that even a road construction programme 50 per cent larger than *Roads for Prosperity* would still not be large enough to outpace the growth of traffic – a finding used in the report to argue for an even greater road programme, but in the event having exactly the opposite effect, of casting doubt on the viability of such a strategy as a whole;[9]
- the Royal Commission on Environmental Pollution's *Eighteenth Report* in 1994 which demonstrated the environmental unsustainability of current trends;[10]
- the SACTRA report *Trunk Roads and the Generation of Traffic*, also in 1994, which demonstrated that increases in road capacity, in conditions of congestion, typically led to some increase in the total volume of traffic, which reduced the duration of any relief from congestion;[11]
- the RAC Foundation's 1995 report, *Car Dependence*, which persuaded the RAC (and some other agencies) to support policies intended to reduce car dependence and envisage the possibility of replacing car use for about 20 per cent of the current car journeys;[12]
- DETR and London Transport research on *Traffic Impacts of Highway Capacity Reductions* in 1998 which demonstrated that road capacity could be reallocated to pedestrianization, public transport priority, etc, without necessarily causing intolerable extra congestion on alternative routes);[13]
- the 1999 SACTRA report, *Transport and the Economy*, which argued that in conditions of imperfect competition, some of the economic impacts of transport investment would not be captured in current appraisal methods – but in some circumstances, road building could make economic efficiency worse, and price increases could make it better.[14]

The picture given is therefore of an evolution of thinking over the whole of the 1990s, in which *A New Deal for Transport* stands as an extremely important statement of a government position, articulating and drawing together the experience of the decade, but not of itself being a new departure. No official publication is ever completely unambiguous, but in the White Paper there was set out in uncommonly clear language the strategy which had been coalescing since the abandonment of the 1989 road policy, *Roads for Prosperity*.

At the heart of this was a key concept, 'new realism', proceeding from a recognition that there is no feasible road construction programme that would be capable of keeping pace with unrestricted traffic growth – a recognition which was swiftest in cities, and embraced politicians of all political parties, and transport professionals of all disciplines. This implies that 'predict and provide' *cannot* be successfully implemented, and it follows that a range of alternative approaches, with demand management as an essential part, was forced on to the policy agenda for reasons of arithmetic as much as ideology.

In the resulting package, there is an emphasis on the various tools of demand management from pricing to physical restraint, a re-emphasis on public transport as an expanding rather than contracting industry, a rediscovery of short distance local movement especially by foot and cycle, and an appreciation that this implies land-use planning which encourages such patterns of movement. The balance among these aspects varies, as also does the relative support for different instruments (for example, traffic bans versus pricing), and between different sorts of public transport (such as bus or light rail), and between different claims for the reallocation of road space (bus lanes, lorry lanes, cycle lanes, pedestrians), and between management of the existing road network or some expansions which – though always on a lower level than previously – are sometimes seen as almost entirely inconsistent with the package, or a contingent way of easing its implementation.

Such a package may in general (though not in every particular) be supported in at least four different ways: efficiency of movement, reducing local and global environmental damage, avoiding divisive and politically damaging protests and demonstrations, and costing less. So a rather large constituency of professional and political support was possible. Hence the 'New Deal for Transport' should be seen as the statement towards which government thinking had been evolving, under both Conservative and Labour administrations, recognising a nonparty policy imperative. If there had not been a change in power in 1997, there would still have been something rather similar in content (albeit possibly different in language) to the New Deal.

Yet the chapters in this book record many examples where, to say the least, it is rather difficult to interpret current transport statements as the result of a seamless evolution from 1997 principles to twenty-first-century practice. Rather, some inconsistencies asserted themselves more and more strongly to the extent that the rather elaborate shift in strategic approach that had been articulated and then become entrenched through the 1990s, was finally abandoned in late 2002.

Has There Been a U-turn? From the White Paper to the Crisis of 2002

Bad luck?

Although it did not go as far as some commentators had hoped, in retrospect it was remarkable how well the 1998 White Paper was received, in the media and in public opinion. Even allowing for a honeymoon mood enjoyed, for a while, by the new government, there was a widespread acceptance of its radical language and its basic analysis: only a few voices said the approach was wrong, and even fewer doubted the government's intention to deliver it.

By contrast, the field of transport is now reckoned to be one of the least successful areas of government achievement – including by the government itself, in suitably rounded phrases like 'well, of course we would be the first to say we need to do better'. In part, this must be due to events that cannot be described as fundamental faults of a transport strategy, though they became symbolic of a general discontent – examples are the Hatfield rail disaster, the tactless stupidity of the 11 September e-mail and the peculiar combination of 'countryside' issues that somehow combined fox hunting with fuel taxation.[15] It was not inevitable that London, the one city most interested in making early use of the charging powers given to it by the government, should be led by a Mayor who was so closely seen as a voice of opposition to New Labour. Railtrack's fall, or assisted fall, was a product of a privatization and semi-deprivatization that had little directly to do with a sustainable transport policy at all. None of these developments was greatly enjoyed by transport planners, but essentially they neither prove nor disprove the viability of the White Paper's strategy, nor the government's commitment to pursuing it. Probably more important evidence is seen in important shifts in two areas, the *10-Year Plan* and ambiguities in the implementation of its multi-modal studies (MMSs).

The 10-Year Plan

Early in 2000 there was a gentle opening shot when the government, in the context of its first annual report required by the Road Traffic Reduction

(National Targets) Act (1998), announced its intention to focus not on traffic reduction but on reducing the negative effects of traffic growth. Later in the year, the *10-Year Plan* was published, intending to do two things, which have turned out not to be entirely consistent, or at all easy. These were, first, to convert the promises and aspirations of the White Paper into deliverable – and more important, *delivered* – improvements, to counter an increasing public and media criticism that in spite of all the fine promises, nothing much was happening; and second, to counter what the government saw as a perception of being 'anti-car' which was, or would be, a threat to its electoral base.

From the beginning there were professional doubts about this plan, but they were muted, because of the sheer pleasure of actually having a government which would talk about planning with a ten-year horizon. But that was a short-lived fool's paradise, partly because it was rapidly overwhelmed, as before, by the immediacy of the short term headlines: Hatfield, the fuel tax protests, the fall of Railtrack. And partly because internal, technical, inconsistencies will always, in the end, reassert themselves. The problem was that (for reasons not pursued here) a bizarre method of measuring the 'percentage change in congestion' was adopted which had the property of turning invisibly small changes in travel time into worthwhile-sounding changes in congestion (Table 10.1) While this issue itself was of interest only to a small minority even of transport professionals, its discovery underlay a loss of confidence in the role of the *10-Year Plan* as a viable plan for delivering year by year detectable improvements. An early intention to produce high profile annual reports on 'progress so far', comparing the targets with – it was assumed – noticeable movement towards them, was downplayed.

In May 2002, the House of Commons Transport Committee published a highly critical report on the *10-Year Plan* (following which the Secretary of State resigned, though not only for that reason).[16] Its argument, in summary, went like this:

- the *Plan* has failed to tackle the increasing cost of public transport and the falling cost of car use, which would – *must* – run counter to the longer term objectives of shifting in the opposite direction;
- it backed away from any significant restraint on car use;
- it focused on congestion – and a peculiarly limited definition of congestion at that – at the expense of improving local access to facilities, urban regeneration, safety, social exclusion, health and quality of life;
- even the focus on congestion gave too much attention to capital infrastructure and not enough to operations, management, finance, education and other supportive measures, with the result that the Plan – even if it could have been fully delivered, which was doubted – would be, the

Table 10.1   *10 Year Plan* traffic and congestion estimates.

	Total (%)	Big cities (%)	Interurban trunk (%)
Traffic	+17	+10	+26
Congestion	−6	−8	−5

Source: Department of the Environment, Transport and the Regions (2000) *Transport 2010: the 10 Year Plan.* DETR, London.

Committee said, in complete contradiction to the government's aims to promote equity and social inclusion;

- the proposed action in the *Plan* had no serious time scale and detail against which delivery could be monitored and assessed.

The Committee did not divide on party lines about this main line of argument, and in my reading the criticisms were, broadly, warranted, and were consistent with the Conservatives' policy development in the 1990s *and* with the underlying rationale of the White Paper.

In July 2002, after much discussion and redrafting, a group of 28 Professors with 'transport' in their titles (from engineering, economics, psychology and other disciplines) sent a letter to the new Secretary of State.[17] They said that even if politicians would *like* to be advised that the comfortable combination of selective road building and improvements to alternative methods of transport would improve travel conditions without the need for traffic restraint, 'the evidence is that if traffic growth continues at the rates of recent decades, such a package will not in practice achieve its intended effects'. It went on to argue:

> Investment in public transport infrastructure, and provisions for walking and cycling, are indeed necessary, but in congested conditions they will also need priority allocation of road space, without which a genuinely attractive service will not be possible.

The interesting thing about that sentence was that it was not controversial among the potential signatories. All were happy to sign up, with little if any dispute over wording. That would have been unthinkable ten, or possibly even five, years ago. And the next passage was quite revealing:

> We have a range of different views about the scale of road building that should be undertaken – some of us advocating more, and others less, than is currently planned. But we all agree that efficient road planning depends strongly on a clear understanding that there will have to be active policy intervention to

manage the demand for road space at congested times and places. Without this, the benefits of any infrastructure expansion would be substantially eroded by extra traffic, disappointing car drivers and non-drivers alike.

The important point here is that it would have been quite impossible to develop a significantly large constituency without that first clause. Some would have signed happily for a moratorium on new construction, or close to it; others would like to see a fairly substantial road building programme, and there was no strong sign of a professional agreement on how much.

Recent contributions to this debate had a strange and unexpected resonance.[18] An RAC Foundation report, *Motoring Towards 2050*, argued that technical advances would solve environmental problems largely by vehicle design and fuels, but congestion would remain as a problem and for this there was a clear policy choice: massive road building, or massive increases in motoring costs, or a combination of both. (The RAC Foundation favoured the combination – charging modest prices for the use of roads and expanding the network.) The Foundation then commissioned consultants to explore the high road building option, that is, what sort of road expansion would be necessary to avoid mounting congestion *without* having to resort to pricing?

The answer was that a trunk-road building programme some 50 per cent bigger than promised in the *10-Year Plan*, sustained for 30 years, would just about hold congestion levels constant on the trunk roads, though not counting the disruptive effect of the road works themselves, and not counting the consequent congestion effect on the nontrunk roads which would not be expanded at such a pace. The RAC Foundation issued a press release saying such a road programme would be a well worth-while policy, though the authors actually made it clear that they were making no such recommendation, and the RAC Foundation swiftly explained that, in fact, it wasn't either. But the overall effect was a time-loop back to the mid 1990s: the British Road Federation study had called for a level of construction '50% higher than the Government's road plans', but this, then as now, would still not be enough to reduce congestion.

The multi-modal studies – a problem of disintegration

Initially a parallel and somewhat separate process, the programme of MMSs was being squeezed by the same inconsistencies. These MMSs were a series of over twenty substantial technical studies, carried out by consultants with government support and regional guidance, into the transport problems of a number of designated areas, mostly defined by interurban or peri-urban motorway corridors, but also including some urban areas and conurbations (Chapter 4). The terms of reference were to take road congestion as the starting point, but define the 'nature of the problems' much more broadly than this, and investigate solutions *not* confined to road construction, but to include action on all modes of transport, and the role of charging or other demand management.

Fifteen studies were complete by late 2002, and they show a wide range of approaches and assumptions, with some inconsistencies. But the interesting and disturbing development was shown most clearly in those studies that expressed a genuine attempt to develop multi-modal, coherent, internally consistent solutions for their area. Typically, they used strong language to argue that their recommendations were a 'package' that could not be unpacked. Each element would only work on condition that the other elements were also implemented. Overall, recommendations for rail investment were substantially greater than recommendations for road investment.

But the subsequent treatment of the packages almost completely invalidates such caveats. The recommendations were unpacked, turned into lists of projects, and then the projects referred to separate implementation agencies for further scrutiny, testing and decision. Thus the road recommendations were referred to the Highways Agency – which mostly welcomed them – and the rail projects referred to the Strategic Rail Authority (SRA) – which has rejected, or been cool, about many of them, partly because of absence of funding and partly because they do not easily meet the SRA's own strategic objectives in which road congestion is a small element. A much larger element is the continuing and problematic story of franchises, delays, fines, operating difficulties and blame-shedding in the rail industry. A series of announcements by the SRA at the end of 2002 effectively stated that the programme of massive improvements by 2010, to accommodate a 50 per cent increase in demand, could not be funded (Chapter 5). Rather, as in the days of British Rail (BR), fare increases would have to be used to discourage traffic growth at the busiest times.

Admission of defeat, Christmas 2002

Alastair Darling, the safe pair of hands appointed as Secretary of State for Transport after the ill-starred Stephen Byers, announced that several of the major road construction schemes recommended in the MMSs would go ahead, albeit not in the context of the coherent 'integrated' package of other measures as designed in the studies. At the time, it seemed a half-hearted and muted announcement, with proforma objections from environmental campaigners, but overshadowed by other events. Within a week, leaks started to appear of the content of his departmental progress report, published just before Christmas 2002. This was honest and remarkably clear. It said, simply, what critics had been saying for the previous two years, that the *10-Year Plan*, even including the newly announced road schemes, would not be able to bring congestion down in the decade.

In evidence to the Transport Select Committee, the Secretary of State said that 'it is not the Government's objective to reduce traffic growth . . . the Government's objective is to reduce congestion'. But the revised forecasts said that congestion, and indeed carbon dioxide emissions, would get worse each year during the whole forecasting horizon until 2010, and probably thereafter. The apparent prospect now, therefore, is almost a caricature of unsustainability and unpopularity. The roads programme will face an uphill battle to get the public support to progress construction at the planned rate; if it gets the support, there will be a decade of disruption due to the effects of construction itself, and on completion any relief will be short lived as traffic growth erodes its benefits. There is hardly any practical possibility of delivering the rail improvements recommended in the MMSs, and targets for growth in passengers and freight – while achievable in principle – are suspect. The government feels unable to make any commitment to charging (or other forms of demand management) on trunk roads, which in turn weakens the vigour of its support for local authorities trying to implement those policies in cities. Christian Wolmar, a veteran transport journalist, argues that it is transport that will bring down the Prime Minister.

Where Now?

So where, one might ask, is the justification for the word 'towards' in the title of this chapter? To my mind, it resides in two developments. First, developments in cities have some degree of autonomy, and variety, which enables demonstration of practical and worthwhile experience. Secondly, there is a fundamental political difference between the *10-Year Plan*'s claims that things would get better (which were probably not well-founded), and the December 2002 progress report's revision that they would get worse. As long as the *Plan* could be presented as being capable of success, it diverted attention from its weaknesses. The revision makes that impossible. It is true that ministers continue to struggle with the rather complex syntax necessary to describe deterioration as an improvement – 'the levels of congestion in this country will be significantly less than they would have been, but we will have to try an awful lot harder if we want to reduce congestion' – but meaning does keep breaking through the grammar. The proposition 'vote for us, we will make things worse more slowly than the other lot' is simply not a *politically* sustainable position, and the policy dynamic it initiates can open up possibilities that were previously denied.

There are developments which one can confidently assert will develop during the next period. The first is that voices will be raised, and *not* be effective, for an even greater road building programme. One cannot discount the possibility that for a period such ideas might find a place in the programme of a government or an opposition, but one can discount the probability that they will last long enough to be implemented. Whatever the momentary balance of argument, there simply will not be the sustained conviction, funding, momentum and breadth of support necessary to deliver such a programme. Even if that were possible for a national trunk network, the fault line with the much larger local networks cannot be bridged.

The second, therefore, is that unmanaged traffic growth would proceed faster than the growth in road capacity, and the prospect offered is one of congestion getting worse. This takes away the electoral base which otherwise might be available to weigh growth in provision for personal mobility more highly than environmental damage. Such an electoral base can exist if the growth in mobility is being attractively provided for: it cannot if the year by year experience – and the forecasts – both assert deterioration. This does not take away the underlying personal and social pressures for the growth in personal movement. In turn that must reassert the underlying logic for demand management just as strongly as it did in the mid-1990s.

Thus it is not an accident that the experience of the MMSs is putting unexpectedly intense pressure on the government to come off the fence about its plans for charging. The proposition 'no nationwide road charging before 2011' is entirely defensible as a statement for the next few years, but its careful ambiguity – does that mean 'charging after 2011' or 'no charging ever'? – is not. Any road scheme, or major transport initiative, to be implemented *within* the period of the *10-Year Plan* will live for most of its life outside that period, and it is a matter of elementary appraisal that the scheme design, scale and type of building, size, and even location of a transport investment will be different according to whether this lifetime is one of a priced network, or an unpriced network; and if the latter, unpriced in the context of traffic growth restricted by other policies, or by capacity constraints, or unrestricted due to the return of predict and provide.

An early decision on this is clearly necessary. The government has been entirely open about its distaste for being forced to the point of such a clear statement of intent on the matter, but it is difficult to see how events will allow the question to be indefinitely postponed. 'Plan A' in 1998 was for active encouragement by the government for local authorities to make use of the charging powers they now have, and an active process of preparation by the government to implement a related scheme on parts of the national network. This was on the agenda because there was no other way apparent of improving travel conditions, and does not go off the agenda just because it is problematic. Thus one can see real consideration of a substantial and wide-scale charged network, the prices being set by reference to both congestion and pollution costs as well as direct costs of construction, maintenance and so on, and accompanied by an important range of other improvements as argued elsewhere. In that case, to take a decade of careful and open preparation of public opinion would be entirely sensible. To rush into such a plan too swiftly would be foolish, not radical or responsible. But ministers must make clear that this is their chosen direction, for it is one which pressures will, almost certainly, mount.

Though the logic is powerful, it is not *so* powerful that one can be certain that the current government will take this path, without fear of another government abandoning it half way. So the question of a 'Plan B' must be considered. Essentially, this means demand management and traffic restraint by nonpricing means, mainly physical restriction reinforced by education, public acceptance, planning and improvements to alternative methods of transport. In fact we have considerably more practical experience on how to implement nonpricing based demand management than we do on pricing, and although it is difficult if the pricing incentives are operating in other directions, it is not impossible. In some cases, there even seems to be a greater willingness for public opinion to accept the more radical restraint in (for example) town centre pedestrianization, than the more moderate behavioural change involved in pricing.

At a local level, this is a robust strategy, and indeed the logic is to see a greater emphasis on the local and less on flagship strategic projects and policies planned and delivered at the national level. There have been some good and successful initiatives already, by local authorities earnestly implementing what they saw as the new agenda following *A New Deal for Transport*, with increasing concern and demoralization as they perceived support from the government ebbing away as its policy became increasingly roads orientated. The government's acceptance that such a new orientation (whether accurately perceived or not) would not deliver the goods may have come in the nick of time to offset a feeling of local confusion, even despair, about what the apparent new strategy meant.

These two strands – demand management (whether by pricing or other means) and a reassertion of the local – are thus both justified in their own right, and part of the logic, albeit not yet clearly expressed, of the government's position. Some road expansion will undoubtedly happen, though there is a long path from an announcement to a ribbon, and it is difficult to see exactly how those schemes which need planning approval will be assessed. Although they will cost a lot of money, and raise a lot of great difficulties in relation to their local environmental and traffic effects, and no doubt be greatly divisive in some areas, they must now be seen as a sideshow in the sense that they do not constitute a coherent approach.

Walking, cycling, buses, land-use planning, 'soft' measures, priority allocation of road space, urban trams, traffic calming and pedestrianization all re-emerge – not surprisingly – as robust elements of a new local emphasis, both with and without pricing. New road construction will have little role to play in the urban transport debate, and will continue to be hotly debated for interurban policies, albeit with a little of the sting taken out of the controversy since *neither* side will be arguing that it can 'solve the problem'. The issue of heavy rail remains problematic because of the still unsolved problem of its institutional structure, and the issue of air travel remains problematic because of the mind shift necessary to come to terms with long term sustainability (Chapters 5 and 9). If charging in London sustains its early promise, and is taken up elsewhere, it is clear that the entire equation changes, and one could see a policy trajectory cascading from the local to the national very much more smoothly than was believed. And if it does not, then nonpricing measures will be relied on even more heavily, a 'Plan B' that – though in my view less powerful than 'Plan A' – has its own advantages which are very substantial.

So perhaps the concern about a U-turn is not the central problem. The problem is not that the government has abandoned its sustainable transport strategy and is heading off firmly and with confidence towards an unsustainable one. Rather, in the light of history, the experience is one of having largely wasted two years at the national level, in a diversion which has proved to be a blind alley, and was no strategy at all. This is now nearly, but not quite, recognized by ministers. That there has been a loss of momentum and direction is indisputable. But meanwhile, understanding about sustainable transport strategies at a local level has moved on, and it is time to re-emphasize their critical importance. The challenge for Labour will be to learn the lessons of the last few years and use the innovation and leadership developing at the local level to reconstruct a national strategy for sustainable transport.

NOTES

1   Department of the Environment, Transport and the Regions (1998) *A new deal for transport: better for everyone.* Cmnd 3950, The Stationery Office, London; Department of the Environment, Transport and the Regions (2000) *Transport 2010: the 10–year plan.* DETR, London.
2   Royal Commission on Environmental Pollution (2002) *The environmental effects of civil aircraft in flight.* The Stationery Office, London.
3   European Conference of Ministers of Transport (1990) *Transport policy and the environment.* ECMT/OECD, Paris.
4   Quoted in Goodwin, P (1999) Transformation of transport policy in Great Britain. *Transportation Research Part A*, 33, 661.

5   Department of the Environment (1990) *This common inheritance: Britain's environmental strategy.* Cmnd 1200, HMSO, London.

6   Quoted in Goodwin, P (1999) Transformation of transport policy in Great Britain, 661.

7   Quoted in Goodwin, P (1999) Transformation of transport policy in Great Britain, 662.

8   Department of Transport (1989) *Roads for prosperity.* Cmnd 693, HMSO, London.

9   Centre for Economics and Business Research (1994) *Roads and jobs.* British Road Federation, London.

10 Royal Commission on Environmental Pollution (1994) *Eighteenth report. Transport and the environment.* Cmnd 2674, HMSO, London.

11 Standing Advisory Committee on Trunk Road Assessment (1994) *Trunk roads and the generation of traffic.* HMSO, London.

12 RAC (1995) *Car dependence.* RAC Foundation for Motoring and the Environment, London.

13 Cairns, S; Hass-Klau, C and Goodwin, P (1998) *Traffic impact of highway capacity reductions: assessment of the evidence.* Landor, London.

14 Standing Advisory Committee on Trunk Road Assessment (1999) *Transport and the economy.* The Stationery Office, London.

15 Jo Moore, the Special Advisor to Stephen Byers, then Secretary of State for Transport, sent an e-mail to colleagues in the Department of Transport, Local Government and the Regions at 2.55pm on 11 September 2001 suggesting that the events unfolding in New York could be used as cover for potentially unpalatable government announcements. She wrote, 'it is now a very good day to get out anything we want to bury'.

16 House of Commons (2002) Session 2001–2002, HC 558–I, 27 May. The Stationery Office, London.

17 Transport Planning Society (2002) *Letter of 28 transport professors to Secretary of State for Transport.* www.tps.org.uk (accessed 14 December 2002).

18 Bayliss, D and Wood, A (2002) *Means to mitigate effects of increasing strategic road capacity in line with demand.* RAC Foundation, London; RAC (2002) *Motoring towards 2050.* RAC Foundation for Motoring and the Environment, London.

Index